MW01517715

THE EVOLUTION FROM HORSE TO AUTOMOBILE

THE EVOLUTION FROM HORSE TO AUTOMOBILE

A Comparative International Study

Imes Chiu

CAMBRIA
PRESS

AMHERST, NEW YORK

Requests for permission should be directed to:
permissions@cambriapress.com, or mailed to:
Cambria Press
20 Northpointe Parkway, Suite 188
Amherst, NY 14228

Library of Congress Cataloging-in-Publication Data

Chiu, Imes.
    The evolution from horse to automobile: a comparative international study / by Imes Chiu.
        p. cm.
    Includes bibliographical references and index.
    ISBN 978-1-60497-546-8 (alk. paper)
    1. Automobiles—Technological innovations—History—20th century. 2. New products—Social aspects—History—20th century—Case studies. 3. Technology transfer—History—20th century—Case studies. 4. Transportation, Automotive—Social aspects—History—20th century—Case studies. 5. Jeep automobile—History—20th century. 6. Horsemanship—History—20th century. 7. Displacement (Psychology) I. Title.

    TL15.C426 2008
    303.48'32—dc22

2008025435

For my father, Chiu Cheng Beng (Francisco Chiu)

TABLE OF CONTENTS

LIST OF TABLE AND FIGURES

ACKNOWLEDGMENTS

Three things made this book happen: a Japanese proverb, a group of friends, and a lucky break.

The words 七転八起 (fall down seven times, stand up eight) sustained my writing.

Kristen Ebert-Wagner's expertise in the English language and software applications made this book possible. She brought out the best of my ideas and made them *shine.* She took away many burdens—formatting manuscripts, meticulously checking compliance in notation and style, indefatigably tracking people down to obtain permission letters, ensuring a complete and accurate bibliography, and even detecting and resolving problematic references.

Every single chapter in this book began and ended with the encouragement of Dolina W. Millar. To her, I give my most heartfelt thanks for her unfailing support, wonderful history lessons, enjoyable conversations, *meticulous* editing, and pick-me-up sticky rice. Pinky gave her time, ideas, and editing expertise without any thought of return. She was there throughout the entire journey. I feel so fortunate to be counted as one of her friends.

I am deeply grateful to Cambria Press. In particular, I thank Dr. Paul Richardson for believing in my work. Dr. Richardson gave invaluable advice and knew how to hearten a writer's soul. I also thank Toni Tan for her help in leading this work to publication. Thank you both for my lucky break. Finally, thanks go to Sharon Berger for her help during the final stages of publication.

Mark Shepard, thank you for providing a safe haven for creative ideas. Margie Towery's indexing guidance was invaluable. She taught me with precision and efficiency. I am grateful for her kindness and generosity. Special thanks to Terri Hudoba and Kara West for their help in the final stages of the book.

I thank Michael E. Lynch, Ronald R. Kline, and Trevor J. Pinch for their helpful comments on this work. My special thanks to Stephen H. Hilgartner for his kindness and his comments, which encouraged me to expand this project. Any weaknesses in argument or misinterpretations of scholarly work are entirely mine.

The many wonderful images, in both pictures and words, were made possible by Jacqueline Côté-Sherman; Bill Spears; Norm Booth; Elizabeth Bender at Feld Entertainment, Inc., Vienna, Virginia; Erin Foley at the Circus World Museum, Baraboo, Wisconsin; Joe Mischka at Heart Prairie Press/Mischka Press, Cedar Rapids, Iowa; Paul Carnahan and Marjorie Strong at the Vermont Historical Society, Barre, Vermont; Tracy Powell at *Automobile Quarterly*, New Albany, Indiana; Aloma Harris and Jonathan M. Gordon, Esq., at Weston Benshoof Rochefort Rubalcava MacCuish LLP, Los Angeles, California, on behalf of the William H. Mauldin Estate; and Dan Curnow and Schoolmaster Robert Weaver at the Crossroads School for reprints of Trudl Dubsky Zipper's watercolors.

I am grateful for the assistance of Arlene Pasciolla of The Reader's Digest Association, Inc.; Heidi Reuter Lloyd of RDA Milwaukee; and Trudi Bellin of Reiman Publications, who together led me to the Circus World Museum. I am indebted to Kristine Withers of the U.S. Cavalry Association, who in her generosity of time and skill filled in many missing pieces in the second case study. I also thank, for their warm support,

Professor Emmanuel Torres, and Mrs. Gilda Cordero Fernando, owner and publisher of GCF Books.

The archival materials in this book largely came from Cornell University libraries such as the Johnson Management Library, the Engineering Library, Kroch Library, and the Fine Arts Library. I am most thankful to Cornell's Olin Library and its many offices, particularly the wonderfully helpful and capable reference librarians. The Cornell Library Annex, Library Acquisitions, and the Interlibrary Loan Office were indispensable in this research project. I also thank the reference librarians at the Monroe County Public Library System. The help of Tokyo University, Ateneo de Manila University, and University of the Philippines is much appreciated.

I am most grateful to Virginia Cole of Cornell's Olin Library, whose library expertise found many rare primary sources; Fred Kotas at the East Asian section of the Kroch Library; David Brumberg of Olin Library Acquisitions; and intellectual property officer Peter Hirtle; Cornell Veterinary School's Judy L. Urban, Ann Townsend-Poors, and Margie Vail. My special thanks to Kiyoshi Shiraishi, Yashushi Uchida, and Dan Plafcan for their assistance in Japan.

For ideas, support, and inspiring conversations, thank you to Daina Taimina, David Henderson, Boris Dzikovski, Robert Ackerson, Patricia Garnier, Samba Sow, William Sow, Fanta Sow, Ann Leonard, Victoria Ying, William Harold Newman, Margaret "Sugar" Franklin, Manjari Mahajan, Mahnaz Mousavi, Lindy Feigenbaum, Claire Lyngå, Kurt Karlsson, Saga Karlsson, Andrew Webster, Jonalee and Ben, Vivian and Robert Orshaw, Yvonne Hoehn, Victoria Rodriguez, Marjorie Mudrick, Winnie So, Bernie Que, Merritt Crocker, Kim and Mike Thies, Judy Radabaugh, EunYun Park, Kira and Peter, Astrid and John October, Clare Pollard, Ulla Uusitalo, Barb Bridges, Beverly Bradley, and my wonderful girls at AXO.

For prodding and praise, I thank my family, Manong Manolo, Maribelle for her clever advice and tireless encouragement, Francis and Bernice, Michael for photographs of the jeepney, Lilibeth, Lance, Liam, Luc, Leon, and my caring brother Carlos, his wife Carol, and little Claire.

Finally, my special thanks to my parents, Anita Lim and Francisco Chiu, who taught me my first lesson on how to pick myself up.

THE EVOLUTION
FROM HORSE
TO AUTOMOBILE

FIGURE 1. 1905 illustration: Putting the horse in the horseless carriage.

Source. *Life*, January 19, 1905, 68.
Note. Caption reads "A COMPROMISE. By a lover of horses."

FIGURE 2. 2005 jeepney: Putting the horse in modern Philippine transport.

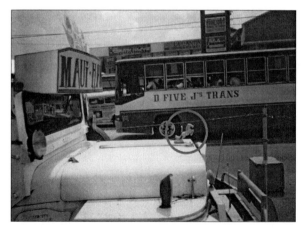

Source. Michael Chua, Michael Agricultural Supply, July 2005.
Note. Horse chrome encircled.

INTRODUCTION

Why would a 2005 form of mass transit in the Philippines carry equine symbolism similar to that depicted in a 1905 American cartoon illustration? What links the symbolic expression of a horse from a hundred years ago with a Philippine vehicle of the present? One does not really think of the horse when driving an automobile. Yet there was a time, in many places for many people, when such was the case. This book revisits that moment when this connection was immediate and palpable.

At the turn of the century, the idea of a machine with a horselike quality—the ability to be self-propelling—seemed preposterous. What seems absurd to us today is the idea of the automobile's dependence on the antiquated horse for acceptance and even survival. We, having largely forgotten our past and, with our early-twentieth-century predecessors, looking ahead to an unknown future, share a similar sense of puzzlement: how could the motorcar be even *remotely* interchangeable with the horse?

Today, the horse may appear at special events but is seen rarely in everyday life. The horse may be bred for sports and entertainment—polo matches, horse racing and betting, rodeos, riding schools—but is not commonly used for everyday work and errands. Most people today do not even know how to ride a horse. Just as we today have trouble envisioning the modern world running on muscle power, people a hundred years ago could not envision a world of horseless carriages: what could possibly substitute for a form of transport that comes home on its own?

At the turn of the twentieth century, the introduction of the motorcar[1] to public roads frequented by horse-drawn carriages caused great public alarm, not only in rural[2] but also in urban areas. This new contraption—a horseless carriage, no less—looked deviant and dangerous.[3] The sight caused such public disturbance that one motorcar owner was arrested for driving to the entrance of Central Park.[4] One study explains that people often do not know what to do with a new technological artifact.[5] Despite price reductions and performance improvements, the motorcar remained suspect to the early-twentieth-century mind.

Why did people reject for so long, at times passionately, what is now considered a sensible option, a panacea for filthy streets and myriad diseases, and a form of transport that ceased to consume when not in motion? A U.S. Army colonel emphatically declared in 1940, "A machine has no life; horses have—that is the radical difference."[6] Indeed, years before the arrival of the Ford Model T, many motorcars were already being advertised to be cheaper than horses; yet this economical alternative failed to spur immediate demand.

Charles E. Duryea, the automotive pioneer who built the first American car in 1893, observed that people "decry rather than ask for"[7] innovations. Yet the standard explanation for automotive ubiquity considers only rational behavior: cars replaced horses because they were much more affordable, more efficient, and cleaner. Is it then simply common sense, albeit following on the heels of initial shock and hysteria, that ultimately accounts for motorization?

This study examines the conversion of users. To understand the motivating factors in mass adoption, I focus on perceptions and practices associated with horses and motorcars in three different settings during three different periods. All three cases begin with the motorcar in the periphery: all three end with it achieving ubiquity. This multiple-case design is used for the purpose of theoretical replication.[8] Results in all three cases show that a contrived likeness to its competitor—the horse—contributed to the motorcar's success. The motorcar absorbed the technical, material, structural, and conceptual resources of the technology it displaced.

ORGANIZATION OF THE BOOK

In chapter one, I review the dominant literature in technology studies and the history of technology related to automobiling. I identify some gaps in the current use of artifacts, social groups, gender, and price as explanatory devices for technological diffusion. I suggest practices—routines used to carry out work and social obligations—as a potential unit of analysis to bridge some of these gaps.

Chapter two covers the first case of this three-part study. I closely examine the marginal use of the motorcar in the United States during the first decade of the twentieth century and discuss strategies used to transfer practices from muscle to motor power. For primary sources, I use physical artifacts and print media, which include commentaries, editorial cartoons, and advertisements. Advertising of this time arguably serves as a mouthpiece for automotive manufacturers[9] and hence provides evidence of the persuasive measures used to generate demand and respond to consumer concerns. Manufacturers' manuals and printed publications are included to show the physical translations of these persuasive measures.

Chapter three, the second case study, focuses on how automobiling was achieved in a highly resistant environment—the U.S. Cavalry after World War I. Here again the motorcar was peripheral, but unlike in the first case, superficial likeness with the horse was insufficient to effect change in the cavalry world. The demise of the horse tradition constituted a fatal blow to the cavalry's identity and core principles as expressed in the 1914 *Cavalry Service Regulations*[10] and the 1916 *Cavalry Drill Regulations*.[11] *The Cavalry Journal*, an internal military publication, provides evidence of the various coping mechanisms employed to keep the horse, which, ironically, led to the conceptualization of its replacement—the jeep.

Chapter four, the final case study, shows how the persistence of local equine practices permeated the form and functionality of the jeep in the Philippines after World War II. Motorization represented the effort of the Filipino people to practically "ingest" a technology whose resemblance to its foreign origins became a mere suggestion. Evidence includes historical accounts, the *Philippine Commission Report* of 1899–1900,[12] physical artifacts such as decorative motifs and symbolic icons, phone interviews conducted in the Ilocano and the Tagalog (Filipino) dialects, and contemporary accounts and reflections of local literary scholars.

This study suggests that technological change, particularly that surrounded by controversy, sometimes involves the counterintuitive measure of keeping certain elements, concepts, practices, and forms "unchanged." A nascent controversial device may benefit from the goodwill and popularity earned by its predecessor by assuming some of its functionality, physical representation, and work routines.

ENDNOTES

1. According to the *Merriam-Webster Collegiate Dictionary*, 11th edition, the word "motorcar," meaning "automobile," dates from about 1890.
2. For discussions on the anti-car crusade, see Ronald R. Kline, *Consumers in the Country: Technology and Social Change in Rural America* (Baltimore, MD: Johns Hopkins University Press, 2000), esp. chapter 2. Also, Ronald R. Kline and Trevor Pinch, "Users as Agents of Technological Change: The Social Construction of the Automobile in the Rural United States," *Technology and Culture* 37, no. 4 (1996): 768–95.
3. Charles E. Duryea, "As It Was in the Beginning," *The Automobile*, January 7, 1909, 47.
4. R. H. Thurston, "The Automobile or Horseless Carriage," *Collier's*, April 28, 1900, 9.
5. George Basalla, *The Evolution of Technology* (Cambridge: Cambridge University Press, 1988), 139.
6. Colonel H. S. Stewart, "Mechanization and Motorization: The Final Chapter Has Not Been Written," *The Cavalry Journal* 49, no. 217 (January–February 1940): 41.
7. Duryea, "As It Was in the Beginning," 47.
8. Robert K. Yin, *Case Study Research: Design and Methods* (Thousand Oaks, CA: Sage Publications, 1994), 46.
9. Pamela Walker Laird, "The Car Without a Single Weakness: Early Automobile Advertising," *Technology and Culture* 37, no. 4 (1996): 797.
10. United States, War Department, Office of the Chief of Staff, *Cavalry Service Regulations, United States Army (experimental), 1914* (Washington, DC: GPO, 1914).
11. United States, War Department, Office of the Chief of Staff, *Cavalry Drill Regulations, 1916*, War Department Document No. 340 (Washington, DC: GPO, 1917).
12. United States, Philippine Commission, 1899–1900, Jacob Gould Schurman, George Dewey, Elwell Stephen Otis, Charles Denby, and Dean C. Worcester, *Report of the Philippine Commission to the President, January 31, 1900[–December 20, 1900]*, 4 vols. (Washington, DC: GPO, 1900, 1901).

CHAPTER 1

CONVERTING CONSUMERS: THE CONCEPTUAL DEPENDENCE OF CONTROVERSIAL ARTIFACTS

The concept of practices, namely, ways of doing things, has not been used extensively to examine technological change and diffusion. Research on technology tends to break down its analysis into two major components—people and things. The dynamics of the relationship between the two have been examined through various lenses—social groups, networks, gender, economics, technical content, and functional features, among others—to explain how an artifact comes to be, or why a certain social order exists. Although studies focusing on people and things complement each other, they differ on many fundamental points, particularly with regard to the agency and impartiality of social actors. Actor-network theorists and gender scholars, for instance, see artifacts as forces that shape social orders, whereas social constructivists recognize agency among humans exclusively.

Little work has been done to explicate the motivational factors of agency, particularly in cases where an artifact initially deemed ineffective or superfluous becomes an everyday necessity, such as the automobile at the turn of the twentieth century. Farmers saw it as a "devil wagon" but later adopted it for use as an all-around device and power source.[1] What makes a social group change its position about a particular artifact? How did the devil wagon overcome its notoriety to become a prosaic mainstream device?

These questions direct the research in this book. While they may have been asked before, my intent is to show a different approach to the problem of newness. Preexisting practices and work routines used as explanatory devices have something interesting to say about diffusion strategies and localization measures.

ARTIFACTS

The general public perception of a new technology as the product of a few geniuses has been questioned by researchers such as George Basalla, who prefers a Darwinian worldview to explain technological change. According to Basalla in *The Evolution of Technology*, "Novel artifacts can only arise from antecedent artifacts [...] new kinds of made things are never pure creations of theory, ingenuity, or fancy."[2] Technological change in this model may be profoundly constrained by the past.

Basalla's theory of technological evolution is characterized by four major concepts: novelty, continuity, diversity, and selection. He defines novelty as variations on old things. Basalla argues that the modern world spins out new artifacts as a matter of routine and compulsion. New artifacts constantly emerge from the old; thus a definitive link exists between new and existing artifacts. Old artifacts never completely disappear, even when new artifacts emerge to replace them. Consequently, technological diversity increases over time. However, certain selection decisions are made about which artifacts are to be fully developed and reproduced.[3]

Basalla argues that there is no universal criterion for functionality. Each society decides on the usefulness of a particular artifact. Local needs determine which things to keep, use, and replicate. This decision is not driven by some universal biological need, such as the human body's need for water. On the contrary, an artifact may be useful to one society but not to another. According to Basalla, "Often it is difficult to determine precisely what is to be done with a new device."[4] For Basalla, things themselves do not force a certain use, despite their possessing an inherent momentum to diversify.

Basalla uses the example of the automobile to illustrate the social construction of technological necessity. The automobile began as a plaything for the rich during its first decade of existence, around 1895–1905.[5] Basalla observes that the development of the automobile was not due to some grave international horse crisis or horse shortage.[6] The motor truck was not a response to some scarcity in horse supply or steam-powered machinery. Rather, "the *invention* of vehicles powered by internal combustion engines gave birth to the *necessity* of motor transportation."[7] In other words, for Basalla things came first, followed by the need for them.

Basalla gives another, more basic example: the wheel. In Mesoamerica from the fourth to the fifteenth centuries, people created miniature wheeled figurines for religious purposes but never put the wheel itself to practical use, even though by this time its mechanical principles were thoroughly understood.[8] Mesoamericans did not find wheels useful for their type of terrain and thus used them mainly for ritual and ceremonial purposes.[9] Similarly, the automobile was initially seen as a nonessential in the United States at the turn of the twentieth century. In both cases, the current local culture determined the efficacy of the artifact, although not necessarily all its features.

Basalla suggests that some aspects of a new technology mimic those of its predecessors despite having no contemporary applications. For example, the traditional cord handles of Congo pottery continue to be replicated in the design of contemporary pottery handles made from clay.[10] He refers to anthropology to describe this phenomenon:

> The regularity with which new materials are handled and worked in imitation of displaced, older ones has led anthropologists to coin a word to designate the phenomenon: skeuomorphism. A skeuomorph is an element of design or structure that serves little or no purpose in the artifact fashioned from the new material but was essential to the object made from the original material.[11]

Basalla does not explore why old components become incorporated into new artifacts, but he does attribute selection decisions to "selecting agents," those "productive individuals capable of making the choices and changes needed to shape the material world as they see fit."[12] For instance, he attributes the diffusion of the gasoline automobile to a group of U.S. Midwesterners who saw an opportunity to exploit their region's natural and industrial resources. The Midwest was rich in hardwood. It was the nation's center for carriage and wagon production, with an infrastructure ready to build the body of the motorcar. It was also home to companies with experience manufacturing stationary gasoline engines. These resources put the Midwest at a competitive advantage for gasoline car production.

Thus selectors, such as these Midwestern businessmen, are an enterprising group with the independence to choose which artifacts are to be mass-produced. Basalla suggests that any artifact could be widely distributed through the sponsorship of this select group:

> The selectors do not represent all segments of society nor are they necessarily concerned with the public's welfare. However, they have the freedom to decide which of the competing novelties would be replicated and incorporated into cultural life.[13]

While Basalla touches on how decisions are made by a select few, it is not clear how or why others would follow their lead. Why would the rest of society accept and even purchase an artifact that a select few have decided to replicate and diffuse? Basalla attributes the motivations of these selectors to socioeconomic rationalist arguments based on profitability and technical superiority (for example, the superior performance of gasoline cars over steam and electric vehicles for long-distance travel, their lower maintenance cost, and so forth) but takes for granted the motivations of the larger society.

Why would the rest of the buying public replace their horses simply because a group of Midwestern businessmen advocated gasoline cars? Is it simply a matter of selectors imposing their will on the rest of society? Social constructivist theories, such as the social construction of technology (SCOT), seek to show more complexity in the process. However, similar to Basalla's view, SCOT also adheres to the idea of technological diffusion as a matter of one group's choice.

SOCIAL GROUPS

SCOT characterizes technological change as a matter of one social group rising above others to impose a standard interpretation on a given artifact.[14] The unit of analysis in this case is social groups. Like Basalla, who described a voluntaristic approach to technological change, proponents of SCOT argue that humans can impose meaning on an artifact. However, unlike Basalla, who believes that *artifacts* bring about other artifacts, proponents of SCOT believe that *social groups* bring about technological diversity.

SCOT recognizes three stages in technological change and diffusion occurring along a process of variation and elimination. The first stage involves the identification of social groups with a stake in the development of a particular artifact; their relevancy is a function of their capacity to influence the artifact's content and form. The second stage, interpretative flexibility, describes how these interpretations conflict with one other. In the final stage, closure and stabilization, one social group's interpretation prevails, and a standard is established.

Within this framework, technological change is characterized by a contest of meanings from which one interpretation emerges to shape the technological form of the artifact. Trevor Pinch sums up the core of technological change according to SCOT:

> The key element is that such groups share a meaning of the artifact— a meaning which can then be used to explain particular developmental paths which the artifact takes.[15]

According to SCOT, these developmental paths diverge primarily because meanings advocated by each social group differ radically from one another. Each meaning generates a different type of technical content.[16] A primary task of the SCOT analyst is to ensure that the groups are homogeneous with respect to the meanings they represent.[17] All members of a group agree on a particular artifact's look and function. Thus, in many ways, one social group represents one worldview. According to Trevor Pinch and Wiebe Bijker,

> We need to have a detailed description of the relevant social groups in order to define better the function of the artifact with respect to each group. Without this, one could not hope to be able to give any explanation of the developmental process.[18]

Here a group-level analysis becomes critical to explaining technological change and diffusion. However, it remains unclear how and why the remaining social groups would concede to the choice of the prevailing group. This question also remains unanswered in Basalla's notion of selecting agents, in which people would conceivably have more things to choose from because of the inherent propensity of technological artifacts to diversify. Why would the remaining groups simply accept the artifact chosen by the "victorious" group? In the final stage in the developmental process according to SCOT, do relevant social groups merge and become one relevant social group relative to the artifact's meaning, or do groups simply dissolve?

Group membership in SCOT tends to be well established throughout the developmental process and, in Basalla's model, fairly exclusive. Basalla's group of selectors do not represent the general populace, while SCOT posits that shared meaning is sufficient to maintain group integrity throughout the developmental cycle. Bijker, however, in his theory on "technological frames" points to the difficulty of establishing a tight one-to-one correspondence between social groups and shared meanings.

Bijker argues that in some instances a meaning proposed by one social group is shared by another. Furthermore, one relevant social group may work with various artifacts simultaneously and interpret each of them in the same way. Hence, meanings and social groupings may not be tightly coupled. For instance, social categories may overlap empirically but differ analytically.[19] Celluloid[20] engineers, for example, may work with a variety of artifacts, each of which they, as a group, interpret differently.[21] Conversely, Bijker also argues, relevant social groups may overlap analytically but remain empirically separate. Again in the case of celluloid, chemists, molders, and pressing-machine designers all represent different sociological categories but share the same analytical frame with regard to celluloid.[22] In this case, people from different occupational groupings interpret a particular artifact in the same way.

In the context of newly emerging artifacts, then, how does a SCOT analyst assign a specific meaning to a specific relevant social group without introducing inconsistencies and overlaps? SCOT criteria are rather straightforward: groupings should revolve around shared meanings of a particular artifact. In practice, however, traditional sociological categories such as gender, age, occupation, economic status, and geographic location tend to be used as organizing categories in grouping members. In an environment of artifacts emerging simultaneously, finding a reliable method to sort people into different relevant social groupings associated with different interpretations becomes a daunting task. Changes in mind-set toward an artifact while it is in development introduce another complication in groupings.

Bijker proposes a theoretical framework that breaks the neat classificatory scheme found in SCOT. "The two sides of analysis," he states, "social groups and technical artifacts," must be folded into "aspects of one world."[23] Theoretical concepts, according to Bijker, should be "as heterogeneous as the actors' activities."[24] But at the same time he proposes a highly restricted model in which actions and interactions are constrained by their technological frame and thus "not everything is possible anymore."[25] Bijker seeks to show that social groups and the meanings associated with them are governed not by their interests in artifacts alone but by some other conceptual factors akin to Thomas Kuhn's notion of paradigm.[26]

USERS

As SCOT became known in various academic circles, its applications moved beyond explaining the development of a certain artifact to include how modifications to a finished product occur. After the early 1980s publication of Pinch and Bijker's article on the safety bicycle,[27] in the mid-1990s Ronald Kline and Trevor Pinch wrote "Users as Agents of Technological Change: The Social Construction of the Automobile in the Rural United States"[28] to show how changes occur beyond the design stage.[29] In their study of the Ford Model T, Kline and Pinch focus on the consumption rather than the design stage, with a consequent emphasis on users rather than manufacturers:

> Furthermore, although manufacturers may have ascribed a particular meaning to the artifact they were not able to control how that artifact was used once it got into the hands of the users. Users precisely as users can embed new meanings into the technology.[30]

Kline and Pinch equate the embedding of new meanings with the finding of new technological applications. Identifying technological change, in this application of the SCOT model, becomes a matter of recognizing new sets of meanings in newfound functionalities.

> New meanings are being given to the car by the new emerging social group of users—in this case, technically competent farm men. To the urban user the car meant transport. For the rural users we have identified, the car, as well as being a form of transport, could be a farm tool, a stationary source of power, part of a domestic technology, or perhaps all of these.[31]

Kline and Pinch suggest that innovation was no longer the exclusive domain of designers and manufacturers. Users themselves could add new features to an artifact's functionality, a state of affairs that suggests an open-endedness in the SCOT model. Indeed, this technological flux is mentioned by Kathleen Jordan and Michael Lynch in their study of the "plasmid prep," a laboratory technique used to insert pieces of DNA into a bacterial medium in order to create genetic material for experimentation.[32] Jordan and Lynch argue that even in a highly formalized, well-established, and seemingly straightforward laboratory procedure such as the plasmid prep there was a "*continual* genesis of incoherence and fragmentation within the relatively settled development of an established technology."[33] Permanent closure and concession were difficult to establish even in an area that appears to be an exemplar of standardization.

Since its publication, the Kline and Pinch study has attracted subsequent research on users of technology, as evidenced by the recent publication of *How Users Matter: The Co-Construction of Users and Technologies*, edited by Nelly Oudshoorn and Trevor Pinch.[34] Oudshoorn and Pinch expand on what Kline and Pinch initially sought to illustrate in the Ford Model T study—the possibility of finding new uses for familiar technologies. In this study, adoption becomes deeply tied to the notion of localization. Users become the new designers of stable technologies.

Oudshoorn and Pinch also point to the recent trend in feminist studies of moving away from a technologically deterministic perspective toward granting users, particularly women, the capacity to change technological use and design. The common perception that women were hapless victims of technology has been modified by the last two decades of scholarship on women as capable users.[35] The extent of users' influence varies depending on their direct control of a particular artifact and their socio-economic conditions. Implicit in this analysis, and perhaps more explicit in some cases, is a sense of struggle emanating from social groups themselves rather than from the meanings they advocate.

Contrary to the approach of impartiality in SCOT, scholars of gender studies, such as Judy Wajcman, give prominence to power relationships—the dominance of men in the technological world and the disadvantaged position of women. Gender becomes a simultaneously organizing and explanatory device. The meanings assigned to each social group tend to be established a priori, not by the features of the artifact but by the analyst. One goal recent gender studies share with SCOT, however, is to debunk the idea of users as passive recipients of technological change.

The notion of users as empowered social groups in recent technology studies may have arisen as a reaction to the traditional approach of crediting sole authority to designers with direct access to technological form. Steve Woolgar, for instance, argues in his study on the microcomputer in the late 1980s that computer manufacturers attempt to "configure the user" with features designed to control the range of user agency. In the view of such designers, form should direct use, and designers should direct users.

> For along with negotiations over who the user might be, comes a set of design (and other) activities which attempt to define and delimit the users' possible actions. By setting parameters for the users' actions, the evolving machine effectively attempts to configure the user.[36]

Woolgar describes designers attempting to manage user agency through machine designs, a concept akin to that of embedded scripts, advocated by actor-network theorists such as Madeline Akrich.[37] Akrich argues that a "technical object defines a framework of action together with the actors and the space in which they are supposed to act."[38] While Bijker describes the limits of social agency in terms of conceptual frameworks such as "technological frames," actor-network theorists attribute user constraints to artifacts. Artifacts are not simply receptacles of meanings. They become active participants, although their prevalence, paradoxically, relies on their being taken for granted, or "black-boxed." Black-box status becomes a prerequisite for technological diffusion.

BLACK BOXES

Bruno Latour, who pioneered actor-network theory with the publication of *Science in Action*, considers nonhuman actors a relevant group.[39] These nonhuman actors, once they are "black boxed," diffuse into "thousands of copies all over the world."[40] For example, Latour traces the history of the diesel engine from concept development through prototype development, product diffusion, and product recall, to its inventor's suicide. Latour describes the engine during the diffusion stage as "incorporated as an unproblematic element in factories, ships and lorries."[41] Diffusion, for actor-network theorists, is synonymous with commodification.

According to Latour, if a particular commodity malfunctions, it ceases to be taken for granted because people must find the defect. Mechanics and engineers began opening the black-boxed diesel engines when they failed repeatedly.[42] An artifact may also fail if it is unable to satisfy user needs, another condition that could prompt the opening of black boxes. Kline and Pinch show how technically competent farmers did not merely accept the Model T, but instead modified it in various ways to suit their needs.

In actor-network theory the model of diffusion does not suggest that innovations travel through some inherent force; rather, diffusion becomes a matter of eliminating reasons to open black boxes. Diffusion occurs when "people do not do anything more to the objects, except pass them along, reproduce them, buy them, believe them."[43] In other words, trust accompanies diffusion as long as artifacts perform reliably and problems go away, a condition that resembles the notion of stabilization in the third stage of SCOT.[44] In many ways, the last stage of SCOT represents the culmination of an artifact becoming black-boxed.

Does flawless technical execution, then, become the primary incentive for users to pick up an artifact and use it? Actor-network theory suggests that for adoption to occur, an artifact must enroll the interests of users. As Latour states, if no player takes up the ball in a game of rugby, "it just sits on the grass."[45] Enrollment occurs when various interests and goals become aligned.[46] This enrollment is so critical that, lacking it, an artifact "dies," as Latour seeks to show in his narration of *Aramis*.[47]

Aramis is the story of France's failed attempt to create a rapid personal transit (PRT) system that would have combined the workings of a regular railway train with the personal service of a taxi. Separate train "cars" would pick up passengers on demand. Each train car would monitor its own speed and distance; thus the cars would be physically separate from each other. Each would find the most efficient route for its passengers, bypassing train stations as needed. This method of transportation was designed for people living in the suburbs, where stations are typically smaller and served less frequently by conventional trains.

Several sophisticated technological systems would coordinate intercar linkages. The project spanned almost two decades, from the 1970s into the 1980s, before it was dissolved—hence the "death of Aramis." Latour attributes the death of Aramis to a lack of human sponsors (or, in the parlance of SCOT, a relevant social group).

> Aramis had been fragile from the outset—we all know that; not fragile in just one respect, in one weak link, as with other innovations, but fragile on all points…Here is our mistake, one we all made, the only one we made. You had a hypersensitive project, and you treated it as if you could get it through under its own steam…If the Budget Office can kill Aramis, what should you do, if you really care about it? Impose yourselves on the Budget Office, force it to accept Aramis. You can't do that? Then don't ask Aramis to be capable of doing it on its own. If elected officials from the south Paris region can kill Aramis, what should you do? Make them change their minds, or get other ones elected. You don't think you have the power? Then don't expect Aramis will.[48]

While Latour pioneered the concept of nonhumans as social actors with their own agency, his analysis of the Aramis project pleads for human intervention in the early stages of technological development. Latour subscribes to Basalla's notion that things have inherent momentum, except that for Latour diffusion is contingent on nonaction by users to "open the black box." Basalla argues for a form of material inertia—a compulsion of artifacts to diversify[49]—whereas Latour never separates material agency from the network in which an artifact resides.

If technological diffusion for Latour requires black-box status, technological change becomes a matter of negotiation among human actors. Latour describes technological change as a "process of negotiations between the innovator and potential users" and the manner in which "the results of such negotiations are translated into technological form."[50] These negotiations may be facilitated by what actor-network theorists call "mediators."

According to Akrich, "if we are to describe technical objects, we need mediators to create the links between technical content and user."[51] Unlike in the SCOT model, in which social groups by virtue of their relevancy directly influence an artifact, Akrich describes a mediator who intercedes for users. The same basic idea echoes in the notion of "boundary shifters" described in Trevor Pinch and Frank Trocco's *Analog Days*.[52] Salespeople, for instance, are boundary shifters by virtue of their direct access to both manufacturers and users.[53]

Boundary shifters, according to Pinch, are people who "move from one world to the other," and "apply the knowledge, skill, and experience gained in one world to transform the other."[54] Salespeople bring lessons learned from users back to manufacturers, who in turn modify a particular artifact based on user feedback. Salespeople as mediators are sometimes users themselves, as exemplified by Pinch's story of David van Koevering, who sold synthesizers to rock and roll musicians by capitalizing on his own experience as a player of the synthesizer.

Thus an important point shared by SCOT and actor-network theorists is that users must have some means to transmit their ideas to designers. Mediators provide a communication channel through which social groups can negotiate without altering an artifact themselves. But if such a channel indeed exists and is effectively used, why would the utility of many nascent artifacts not be immediately evident and beneficial to all potential users?

GENDER

Gender studies tend to answer these questions in terms of humans subjugating other humans. Around the time Woolgar published his article on user configuration, Wajcman published *Feminism Confronts Technology*, which proposes that technology is a product of the "distribution of power and resources between different groups in society" and challenges the way social constructivism casts technology as neutral.[55] Although later studies on gender issues deemphasized the theme of disenfranchisement, the idea of technology as a thoroughly male dominion resonated in gender literature for many years.

> As with science, the very language of technology, its symbolism, is masculine. It is not simply a question of acquiring skills, because these skills are embedded in a culture of masculinity that is largely coterminous with the culture of technology. Both at school and in the workplace this culture is incompatible with femininity. Therefore, to enter this world, to learn its language, women have first to forsake their femininity.[56]

Wajcman describes the technological world as imprinted with male agenda. Thus the analyst's task is to understand the different ways in which women are subjugated by or excluded from this world. For example, in her analysis of the diffusion of the automobile, Wajcman argues that transportation has the paradoxical effect of confining women to rather than liberating them from their homes.

> I will argue that the transport system, and in particular the dominance of the car, restricts women's mobility and exacerbates women's confinement to the home and the immediate locality.[57]

Wajcman, citing Langdon Winner's article "Do Artifacts Have Politics?,"[58] uses Winner's view of technology as politicized artifacts, systems, and structures[59] to show how the dependence of American women on public transportation has restricted their access to certain areas, preventing them from taking full advantage of the various economic and social opportunities society offers.

Winner proposes that technical things—the machines, structures, and systems of modern material culture—embody specific forms of power and authority.[60] A technology, as both device and system, contains specific arrangements of power that societies use to enforce discriminative social order. For example, the physical arrangement of a particular technology may systematically promote social inequality, as discussed in Winner's much-cited 1986 case study of the Long Island Bridge.[61] Winner argues that the urban planner Robert Moses in the 1950s deliberately and successfully excluded racial minorities and low-income groups from visiting Jones Beach by designing a bridge under which public buses could not pass.[62]

Wajcman uses Winner's argument to advance her thesis that things themselves are charged with the means to advance male dominion. Technological development viewed through a gendered analytical lens becomes a matter of identifying the different ways in which technology, in whatever form, subjugates women. Writes Wajcman, "Even seemingly innocuous technological forms such as roads and bridges embody and reinforce power relations."[63] Thus artifacts are no longer things to be explained but are, instead, the explanatory variables for social order.

Wajcman attacks the traditional notion that technology means industrial machinery and men. Instead, she argues for a less restrictive definition of technology to include household and other devices used primarily by women.

> I have already argued that the traditional conception of technology is heavily weighted against women. We tend to think about technology in terms of industrial machinery and cars, for example, ignoring other technologies that affect most aspects of everyday life. The very definition of technology, in other words, has a male bias. This emphasis on technologies dominated by men conspires in turn to diminish the significance of women's technologies, such as horticulture, cooking and childcare, and so reproduces the stereotype of women as technologically ignorant and incapable.[64]

Women's technologies, such as "horticulture, cooking and childcare," center on practices rather than things. Gender studies expand the traditional definition of technology beyond mere things to a set of skills and activities. While proponents of gender studies may use the term "technology" more inclusively nowadays, they paint a rather restricted picture: First, the world of technology favors men. Second, to participate in this world, women must become like men. Third, although this world favors men, there are such things as women's technologies, which appear to be associated largely with household-related work.

According to this view, even so-called women's technologies ultimately work to improve the lot of men. The labor saved by early household appliances was that of men, not women. For instance, the cast-iron stove eliminated the cutting, hauling, and splitting of wood, all tasks performed by men, but cooking-related tasks, which were performed by women, remained labor-intensive.[65] Thus, regardless of whether a technology is designed for men or women, its outcome inevitably benefits men, according to Wajcman's thesis.

It is unclear whether in Wajcman's framework it is possible to have a gender-free technology, and if not, how an analyst is to distinguish between men's and women's technologies. If we accept that the culture of masculinity is coterminous with the general culture of technology, it becomes difficult to separate, either analytically or empirically, technology from masculinity.

Judith McGaw sheds some light on this matter by defining feminine technologies as the "tools, skills, and knowledge associated with the female majority."[66] She associates technology not just with things but also with skills and knowledge, which include practices and the uses of things. McGaw argues that focusing on women and technology clarifies the idea that technology is ultimately dominated by men:

> It is also true that until we began to study women and technological change, we were able to remain unaware and ignorant of technology's masculine dimensions—we studied inventors, engineers, and entrepreneurs as though they were simply "people," oblivious to the ramifications of the overwhelming masculine predominance, both numerically and politically, in the so-called technological professions.[67]

McGaw argues along lines similar to Wajcman's in terms of the general disenfranchisement of women, although McGaw attributes the cause to men themselves as social actors rather than to the things they make. McGaw argues that turning a blind eye to gender issues conceals the fundamental characteristic of the social group that dominates technology—men as technicians.

Indeed, the "old-boy network" has been identified by many women as an exclusive club that helps only men and sometimes even blocks women's advancement. John Law's description of successful engineers as "heterogeneous engineers" because of their ability to maneuver physical as well as social relations[68] fails to consider how this ability might be enhanced, in certain cases, by the gender-influenced ability to access those resources, rather than simply to their having well-rounded skills.

Cynthia Cockburn argues that technology may be designed to promote precisely this gender-driven agenda. In her study on the early printing industry, Cockburn shows how male workers advocated for the use of the Linotype typesetter over the Hattersley because the former did not require the use of distribution work typically performed by women or child laborers. Cockburn believes that the heavy lifting required as part of composing work in letterpress printing effectively marginalized women to low-paid finishing jobs such as bookbinding.[69]

Cockburn writes that during the late nineteenth century, some small print shops were run completely by women. Women possessed all of the skills necessary to do composing work and engaged men merely to perform heavy lifting and carrying. However, men influenced the development of printing technology such that the control of the typesetting machine required physical strength that effectively excluded women from composing work. Cockburn writes,

> The bodily strength component of the compositor's craft may be isolated to illustrate the politics involved. Men, having been reared to a bodily advantage, are able to make political and economic use of it by defining into their occupation certain tasks that require the muscle they alone possess, thereby barricading it against women who might be used against them as low-cost alternative workers (and whom for other reasons they may prefer to remain at home).[70]

Printing equipment could have been designed to be smaller and lighter. Since men typically design machinery with men in mind, Cockburn argues, these machines were made to be either too big or too heavy for the average woman. Consequently, women ended up with inferior job positions. The more prestigious, highly paid positions require bodily strength, not just skill, in controlling machines. Cockburn concludes that printing technology became gendered as a result of its discriminative technological form.[71]

On a similar theme, Ruth Oldenziel's study of the Fisher Body Craftman's Guild during the years 1930 to 1968 shows how various institutions marshaled economic and cultural resources to enforce a male technical domain in the design and production of automobiles.[72] Oldenziel argues that the guild socialized boys through various activities such as model-making contests in order to prepare them to become managers and engineers for General Motors. She characterizes the stereotypical relationship between technology, men, and women as follows:

> Men design systems and women use them; men engineer bridges and women cross them; men build cars and women ride in them; in short, a world in which men are considered the active producers and women the passive consumers of technology.[73]

According to Oldenziel, the world of the passive consumer in the Fisher Body advertisements was conveyed with nontechnical, soft, female imagery, such as the parallel made between the contours of the Fisher car and the soft curves of the female body. The world of production, on the other hand, was "technical," "hard," and "male."[74] When men became consumers, they were still cast as knowledgeable producers and builders.[75]

Women, on the other hand, maintained their roles as passive consumers, or what Oldenziel describes as "receivers of what the boys produced."[76] Technological diffusion in this case is explained in terms of women accepting without resistance what men produce. Technological change, furthermore, is completely controlled by men. This general theme of female disenfranchisement can be traced back to Ruth Schwartz Cowan's seminal book *More Work for Mother*, published in 1983.

Cowan shows how tools, even those designed for women, did not deliver the promised benefits. They merely recast the types of work women did, rather than freeing them from work itself. According to Cowan, "Modern labor-saving devices eliminated drudgery, not labor."[77] In her discussion of the automobile, Cowan argues that the American housewife of the 1950s toiled as much as the American housewife of the 1850s. The only difference was location: in the 1850s, she was shackled to the stove; in the 1950s, she was trapped in her car.

> The automobile had become, to the American housewife of middle classes, what the cast-iron stove in the kitchen would have been to her counterpart of 1850—the vehicle through which she did much of her most significant work, and the work locale where she could most often be found.[78]

Unlike contemporary women, nineteenth-century women spent little time shopping and ferrying goods to their homes.[79] Household goods and services, including medical care, were instead delivered to them. Even in rural areas, supplies were purchased through mail-order catalogues and delivered to homes. Retail stores in urban areas provided delivery services and were generally accessible on foot.

However, during the first two decades of the twentieth century, the burden of transporting goods and services shifted from the seller to the buyer.[80] The automobile, according to Cowan, shifted the responsibility of acquiring household goods from men to women.[81] In this case, the car became "an agent of change,"[82] facilitating the shift in household work from production to consumption-related activities. The availability of household appliances and ready-made goods replaced the need for hired help. Nevertheless, the American wife remained harried with work, just of a different kind. Consistent with the tenor of gender studies, the car as a masculine industrial machine in Cowan's analysis worked to benefit the lot of men but not that of women.

Wajcman's analysis of the automobile as a technology that constrains women was inspired by the work of Cowan. Both propose that the automobile ultimately did not benefit women but instead bound them more tightly to their subservient role. Later scholars such as Virginia Scharff agree that even contemporary housewives spend significant amounts of time ferrying children and goods. Household work performed using the car, such as shopping, came to be cast as entertainment rather than a chore, rendering much of modern-day homemaking invisible.[83]

Indeed, automobiles in their early years provided the means for women to conduct commercial and leisure activities outside their homes with greater freedom and less apprehension. Scharff writes,

> As such, it seemed to some women a perfect solution to the prob-
> lem of gaining admission to public life, especially commercial
> and leisure activities, without exposing oneself to the vagaries
> and annoyances of public transportation. It opened up the pos-
> sibility of independent mobility for those who used it. Extending
> that potential to women meant both expanding the private sphere
> into the realm of transportation and, paradoxically, puncturing
> woman's "sphere" by undermining the already strained notion
> that women's place was in the home.[84]

Although Wajcman may have implied that transport technologies were designed to constrict women, Scharff argues that automobiles in their early years extended a woman's sphere and provided the means for her to leave her home without having to forgo privacy and a sense of security. The closed car, for instance, became an extension of a woman's home, like a living room on wheels,[85] where comfort and convenience came to be associated with feminine features.

> Questions of comfort, for the driver and the passengers, lay at the
> heart of the automobile business debate about woman's influence.
> Whenever industry men and male consumers invoked customary
> notions about feminine behavior, they used the terms "comfort"
> and "convenience" to cover a spectrum of meanings, from sober
> concern for safety to lavish luxury.[86]

Scharff argues that automotive manufacturers operated under a gendered assumption that women wanted fluffy features that had nothing to do with automotive performance. Men, on the other hand, were perceived to value practical features such as fuel economy and horsepower. Scharff claims that product differentiation in motorcars occurred because manufacturers thought men and women wanted different things. Gender considerations drove technological diversity.

> Nevertheless, manufacturers tended to associate the qualities of comfort, convenience, and aesthetic appeal with women, while linking power, range, economy, and thrift with men. Women were presumed to be too weak, timid, and fastidious to want to drive noisy, smelly gasoline-powered cars. Thus at first, manufacturers, influenced by Victorian notions of masculinity and femininity, devised a kind of "separate spheres" ideology about automobiles: gas cars were for men, electric cars were for women.[87]

Scharff suggests that this gendered worldview initially fragmented the automotive market. The gender bias of manufacturers resulted in sex-specific advertising campaigns. At the same time, Scharff recognizes buying power as gender-free.

> However, when automotive designers and promoters, acting in part under the influence of cultural imperatives regarding gender, coupled these desirable attributes with the electric's limited power and circumscribed range, they misread their audience. No law of nature dictated that automobiles could not be designed to be comfortable, reliable, handsome, and powerful, qualities that might appeal to men and women alike. And even if automakers continued to insist that males and females had different automotive preferences, a sex-specific promotional strategy made very little business sense in an economy where consumers, male or female, had some choice, and where families buying only one vehicle were likely to have to accommodate male drivers who were presumed to want to go farther and faster than their female counterparts.[88]

The manufacturers' gendered worldview, according to Scharff, was not ultimately economically sustaining, nor did it make business sense. Although she acknowledges that men at this time typically made most major purchasing decisions, she characterizes paying customers, regardless of their gender, as wielding equal purchasing power in the market.

While arguing for the existence of gender division in the marketing campaign of manufacturers, Scharff simultaneously negates this

division by arguing that universal values such as the power of the pock-
etbook and the common desire for comfort and aesthetics ultimately
superseded gender bias. She proceeds to argue that if manufacturers had
incorporated homey features of comfort into the automobile early on,
they would have attracted mainstream buyers sooner:

> Had manufacturers recognized the benefits of providing mobile
> shelter from the beginning of the automotive era, the private auto
> might have made a more rapid transition from "pleasure car" to
> practical means of daily transportation for middle-class workers,
> both those employed outside the home and those who pursued a
> domestic vocation.[89]

To a certain extent, Scharff attributes significant control to automobile
manufacturers, an idea similar to Basalla's notion of selectors. In arguing that their actions, although uninformed, ultimately determined the
timing of the motorcar's diffusion, Scharff inevitably paints a profile of
automotive manufacturers as powerful, albeit mildly ignorant, in their
capacity to control the diffusion of the gasoline automobile. She soft-
ens this categorization by arguing that there were other manufacturers
and businesses who, more in tune with the emerging automotive market,
took early advantage of the economic opportunities active women drivers represented.

> Defenders of women drivers have never been as numerous as
> detractors. Given the potential market that women drivers repre-
> sent, those supporters have, not surprisingly, included many peo-
> ple who produce and sell automobiles and automotive products
> and services.[90]

Thus, on the one hand, Scharff discusses the disenfranchisement of
women due to gender bias among manufacturers; on the other hand, she
qualifies this position by indicating that some manufacturers saw things
differently. Still, she also argues that even others who exhibited gender
bias later changed their thinking: "In an effort to keep up with consumers' changing demands, producers would at once modify their notions of

gender and the machines they made."[91] Do consumer buying power and taste, then, ultimately drive technological change and diffusion?

Scharff argues that women drivers themselves began to break their own gender biases. "As men registered their indifference to the electric, women were demonstrating their own unwillingness to leave long-distance touring and high-speed driving to men."[92] Scharff departs from the traditional gender-studies assumption that men dominate the technological world at all levels. Rather, she argues that economics overpower gender biases:

> The electric car, marketed primarily as a woman's vehicle, provides a striking example of the influence of gender ideology on automotive production. Paradoxically, the electric's failure also illustrates the impossibility of maintaining rigid gender distinctions in motorcar technology at a time when a declining proportion of customers could afford the luxury of his-and-hers automobiles, and where in any case consumers shared certain preferences regardless of sex.[93]

Thus what began as a masculine machine came to incorporate "feminine" features, such as the electric starter, which all users, including men, came to appreciate. Herein lies the contribution of women to the diffusion of the gasoline automobile: the standard automobile became user-friendly for all.

> The self-starter, the device that would replace the crank, offered advantages to all motorists, but nonetheless began its automotive career marketed as a supposedly feminine accessory akin to doors designed to accommodate long skirts.[94]

Technological diffusion in this case became a matter of dissolving what had previously enforced gender stereotypes. Feminine values such as comfort and convenience that were appreciated by men, albeit surreptitiously, became the industry standard.

> Manufacturers who adopted the self-starter made driving easier for all motorists. They also redefined the boundary between men's and women's automotive spheres, no longer identified as the distinction between gas and electric motorcars.[95]

Scharff in many ways sought to show that although manufacturers may have directly controlled how a technological artifact was initially used, they eventually had to yield to consumer demands. Cowan makes a similar observation when she points to the notion of "market acceptance," which advertisers cannot ignore when selling household appliances.[96] She states, "The machine that was 'best' from the point of view of the producer was not necessarily 'best' from the point of view of the consumer."[97] Here Scharff opposes Basalla's notion of selectors who "have the freedom to decide which of the competing novelties would be replicated and incorporated into cultural life."[98] The early manufacturers Scharff profiles were certainly not Basalla's group of selectors: they did not have absolute freedom. They had to provide products people wanted at a price they were willing to pay.

Kline and Pinch point to general socioeconomic factors that facilitated the diffusion of the Model T, such as support from farm leaders; the use of media such as advertisements, editorials, and articles; road improvements; general economic prosperity among farmers; and the affordability and availability of cars as a result of mass production methods.[99] However, it remains unclear how these incentives assuaged the public's dismay at the devil wagon. Why would people spend hard-earned dollars to replace their beloved horses with something that, just a few years earlier, they had fervently despised? In the next section, I focus specifically on the limitations of socioeconomic and technical factors as drivers of change.

HISTORY OF THE AUTOMOBILE

Historians such as Michael Berger and James Flink emphasize the technical and economic advantages of the motorcar over the horse. Berger argues that "the economic argument in favor of the animal had little validity when one considered the time saved, the increase in potential haulage per vehicle, and the reserve power always available with the automobile."[100] According to Flink, "The motorcar was considered cleaner, safer, more reliable, and more economical than the horse. The car promised to be vastly improved and lower in price in the near future, while the expense and liabilities of the horse seemed insurmountable."[101] From a commercial and performance standpoint, Berger and Flink consider the motorcar a commonsense choice.

Flink further states that "the inability of the industry to produce a low cost vehicle in sufficient quantity was all that prevented the rapid disappearance of the horse in American cities."[102] Flink assumes that people were poised to purchase a motorcar as soon as an affordable one became available. He suggests that the diffusion of the motorcar and the subsequent disappearance of the horse depended largely on the automotive industry's ability to produce a low-cost vehicle that supplied consumer demand. However, evidence suggests that affordable motorcars were available well before the Ford Model T.

As early as 1898 an observer insisted that a petroleum canopy cart carrying two passengers priced at $600 brand-new was "infinitely cheaper than horses."[103] Many motor buggies, particularly around the period 1907 to 1908, were advertised to be "cheaper than horses." Secondhand dealers and brokers offering affordable used automobiles were already advertising in several magazines by this time. An observer in 1905 noted that it was comparatively easy to find a secondhand car of "almost any type at a price very much below its original cost, and in many cases at figures that are really absurdly low."[104] That some automobiles were expensive in these early days does not appear to have been a significant barrier to motorization.

Some scholars, such as Clay McShane, emphasize public health over economics, arguing that the organic aspect of horses proved a major disadvantage when compared with motorcars. Horses were vulnerable to disease and death, as exemplified by the Great Epizootic in 1872, which paralyzed the entire city of Boston. Horses were significantly limited in strength and endurance compared with motorized power, particularly in the hauling of freight. They had a shorter life expectancy compared with cars. They ate prodigiously, even when not in use, caused sanitation and pollution problems with their droppings, blocked traffic because of their bulky size, and even caused traffic accidents because they spooked easily.[105] But if the advantages of the automobile were glaringly obvious, particularly in matters regarding public health and safety, why did governmental officials in the first decade of the twentieth century institute strict motor laws to curb its use? For all the good the automobile could potentially deliver, why was it dubbed the "devil wagon"?

William Ogburn proposes that various parts of a modern culture develop at different rates; hence some parts may change more rapidly than others, resulting in what he calls a "cultural lag."[106] The driver of change, or what Ogburn calls material culture, forces another part of the culture, the adaptive culture, to adjust to a new set of conditions.[107] Failure of the adaptive culture to keep pace with change causes strain in the system. According to Ogburn,

> The independent variable may be technological, economic, political, ideological, or anything else. But when the unequal time or degree of change produces a strain on the interconnected parts or is expressed differently when the correlation is lessened, then it is called a cultural lag.[108]

Thus, Ogburn's model would attribute the presence of horses to a "cultural lag" that eventually becomes corrected. While Ogburn does not specifically compare horses to automobiles, he uses the example of automobiles and highways to illustrate his point. He states, although historically inaccurately, that automobiles and highways were two parts within the same culture that were in proper balance in 1910; the automobile was slow and highways were narrow. "The automobile," Ogburn writes, "traveled at not a great rate of speed and could take the turns without too much trouble or danger."[109]

"But as time went on," Ogburn continues, "the automobile, which is called an 'independent variable,' underwent many changes, particularly the engine, which developed speeds capable of sixty, seventy, eighty miles an hour, with brakes that could stop the car relatively quickly...The old highways, the dependent variable, are not adapted to the new automobiles, so that there is a maladjustment between the highways and the automobile."[110] According to Ogburn, the corresponding adjustment of wider and longer roads resulted from the pressure to accommodate advances in automotive performance.

However, well before 1910 cars were already considered too fast and too dangerous—enough so that traffic laws and regulations were instituted to set speed limits. Judges meted out heavy fines and penalties to reckless drivers. Ogburn ignores mediating variables such as governmental intervention through public policies and the manner in which legal controls such as traffic laws address so-called maladjustments. To what extent and at what speed a certain technological artifact is incorporated into society appears to be determined by more than technological factors alone.

Wiebe Bijker and John Law have even argued that technological artifacts themselves do not evolve from some inner scientific or technical logic but rather are shaped by a range of contingencies, such as economic, professional, technical, and political factors.[111] As Bijker and Law ask, "Why did [technologies] *actually* take the form that they did?"[112] What explains the relationships between various elements in a sociotechnological world?

In *The Railway Journey*, Wolfgang Schivelbusch argues, like Basalla, that the old and the new are linked, although he introduces a new consideration: elements in old artifacts effect a sense of familiarity. Schivelbusch argues that British train cars were purposely designed to resemble horse-drawn carriages in order to assure railway passengers, particularly the upper class, of a familiar riding experience despite the change in locomotion. Even though railways were fundamentally different from highway roads,[113] efforts to emulate horse-drawn carriages were intended to negate the unfamiliarity of riding in a radically new form of transport

> The traveling situation of the more privileged classes was entirely different: their carriages looked like coaches mounted on rails. Not only was this design forgetful of the industrial origin and nature of the railroad, it was a literal attempt to repress awareness of them. The compartment, an almost unaltered version of the coach chamber, was designed to reassure the first-class traveler (and, to a lesser degree, the second-class traveler as well) that

> he was still moving along just as he did in his coach, only at less
> expense and greater speed.[114]

The effort to mask the newness of the experience, however, fell short.
The open boxcars for the less privileged travelers brought industrial
progress to their riding experience.[115] The upper class too, despite the
camouflage of train cars, felt "like mere parcels,"[116] mere objects of an
industrial process.[117] They felt "converted from a private individual into
one of a mass public—a mere consumer."[118]

While Ogburn describes cultural lag as the old trying to catch up with
the new, Schivelbusch shows how new technological forms purposely
couch themselves in terms of the old to make new experiences more
palatable. Hence the effort to continue the traditions of horse carriages in
the design of train cars persisted through the years.

> As far as I know, in Europe there were no attempts to create a pas-
> senger car that would be compatible in its form with the modern
> technology of the railroad—i.e., one that would no longer have
> anything to do with the coach-driven compartment.[119]

Would the acceptability of new technologies, then, hinge on some
faint resemblance to their successful predecessors? While Basalla and,
to a certain extent, Schivelbusch note that skeuomorphic elements in
new artifacts serve little or no functional purpose,[120] the effects of such
elements on technological diffusion have not been studied. The replication of old technological forms as a means to enact old experiences may
conceivably help the introduction of a novelty. To what extent, then, do
designs of artifacts affect social order?

David Noble's study of machine tools shows how the purposeful use
of technology to control social structure does not follow a simple cause-
and-effect rule.[121] In his study of programmable machine automation,
Noble argues that upper-management employees of a General Electric
(GE) factory attempted and failed to use technology to reduce their
dependence on workers.[122] The choice to cut metals using a numerical
control (N/C) machine rather than a record playback (R/P) was motivated

by GE management's desire to gain greater control on the shop floor. In the R/P system a machinist, using blueprints, cuts the first model piece, which the automated machine then replicates. In the N/C system the first model piece is cut using mathematical models, a process that circumvents the machinist. A programmer creates in precise mathematical and algorithmic terms the "sight, sound and feel" of an "automatic machinist."[123]

The N/C system, developed through massive financial support from the U.S. Air Force, supposedly eliminated the possibility of blueprints being stolen by subversives and spies at a time when fear of communism was particularly high. The use of the N/C system would in theory give managers greater control over the production process. However, managers found that the new N/C machines still needed skilled machinists to produce a good finished product. The machines and their software programs were not always reliable and still required the presence of skilled workers along the production process. Control over the machinery, despite efforts to circumvent it, remained in the hands of the workers. Noble argues,

> Although the evolution of a technology follows from the social choices that inform it, choices which mirror the social relations of production, it would be an error to assume that in having exposed the choices, we can simply deduce the rest of reality from them.[124]

Indeed, while N/C initially appeared to provide the company with a powerful means to control workforce productivity,[125] management employees later found that machines could not run by themselves but still depended on the skill, initiative, and goodwill of human workers.[126] The effort to impose a new social order using new technology failed because of the machine's inability to completely replicate skilled labor. Hence, preexisting practices from old work routines persisted through the new technological form.

However, some analysts of technology, such as Latour, argue that the lack of consistency and reliability in human labor requires the need for nonhuman instruments—a way of thinking akin to that of GE's upper management. Humans, according to this line of thought, are substandard when it comes to performing certain tasks. In his example of the hinge-pin or door-closer, Latour describes how this nonhuman device, "delegated with human characteristics," made a compliant substitute for a porter. Mechanization in Latour's view becomes a matter of machines being upgraded and reskilled to compensate for humans being displaced and deskilled.[127] Is the machine, then, a "better" human worker?

Hubert L. Dreyfus, in *What Computers Can't Do: A Critique of Artificial Reason*, argues that "machines cannot be like human beings" but that "human beings may become progressively like machines."[128] Dreyfus points to the fundamental difficulty of exhausting all possible human actions and situations in programming a machine to comprehensively mimic human behavior.[129] Pinch also alludes to the failure of machine-like instruments to capture the creativity required in music-making, or the "irredeemable human features of musicianship."[130]

A human mind can easily sort relevant and irrelevant data, whereas a computer must be instructed, in every circumstance, how to treat each input variation; hence the impossibility of having, in Noble's term, an "automatic machinist." Tasks must be greatly simplified, in minute detail, if they are to be mechanized. As Noble argues, efficiency in a machine factory could result only from simplifying the work itself.[131]

Dreyfus argues that if the mind were made to work more like machines rather than vice versa, it would be possible to create an intelligent machine that could substitute for humans.[132] However, the human brain does not function like a machine;[133] thus, incorporating a thinking machine into work routines could potentially result in the propagation of subintelligent human workers rather than superintelligent computers.[134]

Harry Collins subscribes to the same idea as Dreyfus in arguing that human work must be deskilled, or immensely simplified, in order for machines to take over.[135] In contrast to Latour, who argues that machines are reliable substitutes for humans, Collins argues that machines require humans to compensate for their limitations. Indeed, Noble's study on machine-tool automation supports Collins's argument on the need to translate human work into simplified noncreative tasks that machines could replicate without error.

Collins points to the difficulty of replicating even routine human work. For instance, similar to the difficulty of sorting relevant and irrelevant data, work that appears to be simple in fact requires complex programming. An automotive assembler just knows that debris must be removed first before mounting a wheel,[136] whereas computers must be programmed for all possible variations, as Dreyfus points out, in order to account for anything that could possibly go wrong. As Collins writes, "So long as acts can go on without disturbance, machine-like acts could be reproduced."[137]

DOMESTICATION OF TECHNOLOGY

Communication scholars such as Roger Silverstone argue that a technological artifact is situated not in laboratory-like controlled environments but "in multiple domains and in uneven and often contradictory ways."[138] Although Silverstone studies household- rather than production-related technology, he defines technological change in terms of fitting artifacts into spaces within a larger preexisting sociocultural milieu.[139] Feminist scholars have argued for a broader definition of technology that includes household activities. Similarly, Silverstone focuses on technologies of everyday life.[140]

Silverstone argues that technology must be mediated in order to be accepted,[141] which might explain why early European train cars, as Schivelbusch observed, mimicked horse-drawn carriages. Horse-drawn transport was familiar and prevalent, while the new train car disrupted this private and personal way of traveling. The new locomotive needed to effect elements reminiscent of the popular horse-driven culture. As Silverstone argues, domestication involves bringing objects from public into personal spaces.[142]

While this movement from public to personal space seems unilateral, Silverstone also describes an opposite movement, from private to public space, where domestication becomes a tool for mass production.

> One can think of domestication too, as both a process by which we make things our own, subject to our control, imprinted by, and expressive of, our identities; and as a principle of mass consumption in which products are prepared in the public fora of the market.[143]

Silverstone suggests that domestication could be a simultaneously personal and collective expression. Similar to the idea of localization, a foreign artifact must be adapted in order to be widely diffused.

> The domestication of technology refers to the capacity of a social group (a household, a family, but also an organization) to appropriate technological artifacts and delivery systems into its own culture—its own spaces and times, its own aesthetic and its own functioning—to control them, and to render them more or less "invisible" within the daily routines of daily life.[144]

Ultimately, Silverstone's model argues for the immediate need to neutralize newness in technologies—people need to make new things familiar in order to fit them into their daily sociocultural life and functioning. New things are injected with old, familiar elements in order to be absorbed and, hence, domesticated.

CONCLUSION

What domesticated the devil wagon? Understanding why the motorcar was not immediately adopted, despite its affordability and supposedly superior performance to the horse, demands more than economic rationalist arguments. The SCOT model and Basalla's notion of selecting agents describe how a certain artifact comes to be selected by a particular social group. How other groups concede to this decision provides opportunity for further research. The eventual conversion of the public that precedes technological diffusion leaves some unanswered questions. Why would the utility of a supposedly more technically superior artifact not be immediately evident? For that matter, why would a supposedly more advanced machine, such as the European train car, contain antiquated elements from the days of the horse-drawn carriage?

I argue that, similar to the notion of the continuity of artifactual designs in Basalla's theory of technological evolution, some elements are carried over from old to new artifacts. However, unlike Basalla, I suspect these elements are not superfluous. The effort to conjure old experiences through a new technological medium has been observed by Schivelbusch. Train cars were made to look like horse-drawn coaches in order to "repress"[145] the "awareness" of a new transport vehicle, which in turn made the new riding experience less threatening.

Making new technologies familiar resembles Silverstone's notion of achieving invisibility,[146] such that new technologies in essence become part of people's everyday lives. Achieving this "invisibility" may mean achieving, in Latour's parlance, black-box status, or being taken for granted. Once an artifact reaches this state, it is no longer tinkered with, questioned, or even noticed, but is simply accepted and used. In this sense, the new artifact is given its own space, as Silverstone describes, in people's daily routines.[147]

However, I argue that this black-box status requires more than technical reliability; it also demands conceptual consistency. Conceptual consistency requires aligning the artifact's form and functionality with local expectations, socioeconomic infrastructures, and entrenched work practices. These practices include highly disciplined, sequential sets of routines as well as clusters of informal activities.

When an artifactual concept clashes with people's worldview, such as the horselessness of carriages at the turn of the twentieth century, the controversial artifact is visibly out of place and cannot be simply taken for granted. Similarly, when a nascent artifact demands radical changes in ingrained ways of doing and thinking, it tends to be marginalized. If a nascent artifact is to attain ubiquity, it must achieve the paradoxical effect of being invisible by virtue of being part of the expected.

The concept of invisibility has been used by Scharff to describe the taken-for-grantedness of work performed by women using automobiles.[148] The inertia of work routines may shed light on why women, as gender studies have argued, accept as normal the persistent, heavy burden of household-related work despite the large number of labor-saving appliances available. According to social practice, women are expected to labor over household chores. Women seem to have fulfilled these expectations regardless of any change in environment or tools. Thus the newness of modern devices was made familiar, and perhaps even made acceptable, through the persistence of women's domestic workload.

However, the use of practices as the focus of this analysis is not a matter of tracing movements and actions, as described by Siegfried Giedon in *Mechanization Takes Command*.[149] Giedon writes that during the nineteenth century, movement of all kinds—the gait of a horse, the flights of insects, and the pulses of a heartbeat—was rendered in graphic form.[150] Scholars using various devices copied, point by point, the trajectories of human and animal muscle movement. Such analysis focused on the nature of muscle movement rather than its pragmatic use.

As Collins, Noble, and Dreyfus point out, the inexhaustibility of *intelligent* human movement such as work skills cannot be duplicated by machines; hence the need for machines to be sustained by human actions and decisions. This book focuses primarily on practices as ways of doing things, as work and social routines and expectations. Just as gender and communication studies seek to expand the definition of technology to include work practices, this study focuses on preexisting practices manifested through articulations in the designs and operations of a new technological artifact.

In the next three chapters, I hope to provide some insight into the Bijker and Law question, "Why did [technologies] *actually* take the form that they did?"[151] Three cases in three different time periods and settings show that preexisting practices as explanatory devices have something more to say about technological change and diffusion.

ENDNOTES

1. Ronald R. Kline and Trevor Pinch, "Users as Agents of Technological Change: The Social Construction of the Automobile in the Rural United States," *Technology and Culture* 37, no. 4 (1996): 763–795.
2. George Basalla, *The Evolution of Technology* (Cambridge: Cambridge University Press, 1988), vii–viii.
3. Ibid., 135.
4. Ibid., 139.
5. Ibid., 7.
6. Ibid., 6.
7. Ibid., 7.
8. Ibid., 9.
9. Ibid., 8.
10. Ibid., 107.
11. Ibid., 106–107.
12. Ibid., 204.
13. Ibid.
14. Trevor Pinch and Wiebe Bijker, "The Social Construction of Facts and Artifacts: Or How the Sociology of Science and the Sociology of Technology Might Benefit Each Other," in *The Social Construction of Technological Systems*, ed. Wiebe Bijker, Thomas Hughes, and Trevor Pinch, 17–50 (Cambridge, MA: MIT Press, 1989).
15. Trevor Pinch, "The Social Construction of Technology: A Review," in *Technological Change*, ed. Robert Fox (Amsterdam: Harwood, 1996), 24.
16. Pinch and Bijker, "The Social Construction of Facts and Artifacts," 41.
17. Ibid., 32.
18. Ibid., 34.
19. Wiebe Bijker, *Of Bicycles, Bakelites, and Bulbs* (Cambridge, MA: MIT Press, 1995), 194.
20. Celluloid is a highly flammable plastic made to substitute for more expensive organic substances such as ivory and tortoise shell. It is used to make knife handles, billiard balls, and so forth.
21. Bijker, *Of Bicycles, Bakelites, and Bulbs*, 193.
22. Ibid., 192–197.
23. Ibid., 195.
24. Ibid., 15.

25. Ibid., 192.
26. Ibid.
27. Pinch and Bijker, "The Social Construction of Facts and Artifacts."
28. Kline and Pinch, "Users as Agents of Technological Change."
29. Ibid.
30. Ibid., 775.
31. Ibid., 777.
32. Kathleen Jordan and Michael Lynch, "The Sociology of a Genetic Engineering Technique: Ritual and Rationality in the Performance of the 'Plasmid Prep,'" in *The Right Tool for the Job*, ed. Adele Clarke and Joan Fujimura (Princeton, NJ: Princeton University Press, 1992), 79.
33. Ibid., 84.
34. Nelly Oudshoorn and Trevor Pinch, editors, *How Users Matter: The Co-Construction of Users and Technologies* (Cambridge, MA: MIT Press, 2003).
35. Ibid., 4–5.
36. Steve Woolgar, "Configuring the User: The Case of Usability Trials," in *A Sociology of Monsters*, ed. John Law (London: Routledge, 1991), 61.
37. Madeline Akrich, "The De-scription of Technical Objects," in *Shaping Technology/Building Society: Studies in Sociotechnical Change*, ed. Wiebe Bijker and John Law, 205–24 (Cambridge, MA: MIT Press, 1992).
38. Ibid., 208.
39. Bruno Latour, *Science in Action* (Cambridge, MA: Harvard University Press, 1987).
40. Ibid., 105.
41. Ibid.
42. Ibid., 106.
43. Ibid., 133.
44. The later application of SCOT, however, radically departs from actor-network theory when it argues that localization, or the reopening of the Model T, became a motivating factor in its adoption on the farm.
45. Latour, *Science in Action*, 104.
46. See chapter 3 in Latour, *Science in Action*.
47. Bruno Latour, *Aramis, or, The Love of Technology*, trans. Catherine Porter (Cambridge, MA: Harvard University Press, 1996).
48. Latour, *Aramis, or, The Love of Technology*, 291–292.
49. Basalla, *The Evolution of Technology*, vii–viii.
50. Latour, *Science in Action*, 116.
51. Akrich, "The De-scription of Technical Objects," 211.

52. Trevor Pinch and Frank Trocco, *Analog Days* (Cambridge, MA: Harvard University Press, 2002).

53. Trevor Pinch, "Giving Birth to New Users: How the Minimoog Was Sold to Rock and Roll," in Oudshoorn and Pinch, *How Users Matter: The Co-Construction of Users and Technologies*, 248.

54. Ibid., 314.

55. Judy Wajcman, *Feminism Confronts Technology* (University Park: Pennsylvania State University Press, 1991), 162.

56. Ibid., 19.

57. Ibid., 126.

58. Ibid., 131.

59. Although Langdon Winner takes a nuanced view of technological determinism in his *Autonomous Technology: Technics-out-of-Control as a Theme in Political Thought* (Cambridge, MA: MIT Press, 1978), many characterize his later work as technologically deterministic. For details, see Bruce Bimber, "Three Faces of Technological Determinism," in *Does Technology Drive History?*, ed. Merrit Roe Smith and Leo Marx, 79–100 (Cambridge, MA: MIT Press, 1994).

60. Langdon Winner, "Do Artifacts Have Politics?" in *The Whale and the Reactor: A Search for Limits in an Age of High Technology*, ed. Langdon Winner, 19–39 (Chicago: University of Chicago Press, 1986).

61. Steve Woolgar and Geoff Cooper argue in "Do Artifacts Have Ambivalence?" that the Long Island Bridge was in fact passable for public buses. Woolgar and Cooper build on Bernward Joerges' study, which questions the empirical validity supporting Winner's claims (Bernward Joerges, "Do Politics Have Artifacts?" in *Social Studies of Science* 29, no. 3 [1999]: 411–431). Woolgar and Cooper characterize Winner's Long Island Bridge example as an urban legend. For details, see Steve Woolgar and Geoff Cooper, "Do Artifacts Have Ambivalence? Moses' Bridges, Winner's Bridges and other Urban Legends in S&TS," *Social Studies of Science* 29, no. 3 (1999): 433–449.

62. Winner, "Do Artifacts Have Politics?"

63. Wajcman, *Feminism Confronts Technology*, 133.

64. Ibid., 137.

65. Ruth Schwartz Cowan, *More Work for Mother* (New York: Basic Books, Inc., 1983), 61.

66. Judith A. McGaw, "Why Feminine Technologies Matter," in *Gender & Technology*, ed. Nina E. Lerman, Ruth Oldenziel, and Arwen P. Mohun (Baltimore, MD: Johns Hopkins University Press, 2003), 15.

67. Ibid.
68. John Law, "Technology and Heterogeneous Engineering: The Case of the Portuguese Expansion," in *The Social Construction of Technological Systems: New Directions in the Sociology and History of Technology*, ed. W. E. Bijker, T. P. Hughes, and T. J. Pinch (Cambridge, MA: MIT Press, 1987), 111–134.
69. Cynthia Cockburn, "The Material of Male Power," in *The Social Shaping of Technology*, ed. Donald MacKenzie and Judy Wajcman (Buckingham, U.K.: Open University Press, 1999), 181–183.
70. Ibid., 189.
71. Ibid., 194–195.
72. Ruth Oldenziel, "Why Masculine Technologies Matter," in *Gender & Technology*, ed. Nina E. Lerman, Ruth Oldenziel, and Arwen P. Mohun (Baltimore, MD: Johns Hopkins University Press, 2003), 37–71.
73. Ibid., 40–41.
74. Ibid., 41.
75. Ibid., 50.
76. Ibid., 64.
77. Cowan, *More Work for Mother*, 100–101.
78. Ibid., 85.
79. Ibid., 79–84.
80. Ibid., 85.
81. Ibid., 82.
82. Ibid., 83.
83. Virginia Scharff, *Taking the Wheel: Women and the Coming of the Motor Age* (New York: Free Press, 1991), 147.
84. Ibid., 24–25.
85. Cowan, *More Work for Mother*, 125.
86. Scharff, *Taking the Wheel*, 127.
87. Ibid., 36–37.
88. Ibid., 44.
89. Ibid., 123.
90. Ibid., 33.
91. Ibid., 46.
92. Ibid., 42.
93. Ibid., 49–50.
94. Ibid., 58.
95. Ibid., 60.
96. Cowan, *More Work for Mother*, 102.

97. Ibid., 143.

98. Ibid.

99. Kline and Pinch, "Users as Agents of Technological Change," 772–773. Also in Ronald R. Kline, *Consumers in the Country: Technology and Social Change in Rural America* (Baltimore, MD: Johns Hopkins University Press, 2000), 63–64.

100. Michael L. Berger, *The Devil Wagon in God's Country* (Hamden, CT: Archon Books, 1979), 34.

101. James J. Flink, *The Car Culture* (Cambridge, MA: MIT Press, 1975), 35. Also in James J. Flink and the American Council of Learned Societies, *The Automobile Age* (Cambridge, MA: MIT Press, 1988), 138.

102. James J. Flink, *America Adopts the Automobile* (Cambridge, MA: MIT Press, 1970), 53. See also Flink, *The Car Culture*, 35, and Flink, *The Automobile Age*, 138.

103. Henri Dumay, "The Locomotion of the Future," *Collier's*, July 30, 1898, 23. Many promoters of the automobile consistently argued that cars were cheaper than horses. However, many pro-horse groups argued to the contrary.

104. "Cars New and Second-Hand," *The Automobile*, August 3, 1905, 147.

105. Clay McShane, *Down the Asphalt Path: The Automobile and the American City* (New York: Columbia University Press, 1994), 41–56.

106. William F. Ogburn, *Social Change* (New York: Viking Press, 1922), 200–201.

107. Ibid., 211–213.

108. William F. Ogburn, *On Culture and Social Change: Selected Papers.* The Heritage of Sociology (Chicago: University of Chicago Press, 1964), 91.

109. Ibid., 86.

110. Ibid., 86–87.

111. Wiebe E. Bijker and John Law, "General Introduction," in *Shaping Technology/Building Society: Studies in Sociotechnical Change*, ed. Wiebe E. Bijker and John Law (Cambridge, MA: MIT Press, 1994), 3.

112. Ibid.

113. Wolfgang Schivelbusch, *The Railway Journey: The Industrialization of Time and Space in the 19th Century* (Berkeley: University of California Press, 1986), 84.

114. Ibid., 72.

115. Ibid.

116. Ibid., 72–73.

117. Ibid., 73.

118. Ibid., xiv.
119. Ibid., 84.
120. Ibid., 106–107.
121. David Noble, "Social Choice in Machine Design: The Case of Automatically Controlled Machine Tools," in *The Social Shaping of Technology*, ed. Donald MacKenzie and Judy Wajcman (Buckingham, U.K.: Open University Press, 1999), 161–176.
122. David F. Noble, *Forces of Production: A Social History of Industrial Automation* (New York: Alfred A. Knopf, 1984), 248, 265–266.
123. Ibid., 84.
124. Noble, "Social Choice in Machine Design," 172.
125. Noble, *Forces of Production*, 265–266.
126. Ibid., 276.
127. Latour, *Aramis, or, The Love of Technology*, 301.
128. Hubert L. Dreyfus, *What Computers Can't Do: A Critique of Artificial Reason* (New York: Harper & Row, 1972), 192.
129. Ibid., 41.
130. Trevor Pinch and Karin Bijsterveld, "'Should One Applaud?' Breaches and Boundaries in the Reception of New Technology in Music," *Technology and Culture* 44, no. 3 (2003): 557.
131. Noble, *Forces of Production*, 57.
132. Dreyfus, *What Computers Can't Do*, 99.
133. Ibid., 137.
134. Ibid., 192.
135. Harry M. Collins, *Artificial Experts: Social Knowledge and Intelligent Machines* (Cambridge, MA: MIT Press, 1990), 221.
136. Ibid., 35.
137. Ibid., 39.
138. Roger Silverstone, *Media, Technology and Everyday Life in Europe: From Information to Communication* (Aldershot, U.K.: Ashgate Publishing Limited, 2005), 14.
139. Roger Silverstone, *Television and Everyday Life* (London: Routledge, 1994), 81.
140. Silverstone, *Media, Technology and Everyday Life in Europe*, 1.
141. Ibid., 17.
142. Silverstone, *Television and Everyday Life*, 98.
143. Ibid., 174.
144. Ibid., 98.
145. Schivelbusch, *The Railway Journey*, 72.

146. Silverstone, *Television and Everyday Life*, 98.
147. Ibid.
148. Scharff, *Taking the Wheel*, 147.
149. Siegfried Giedon, *Mechanization Takes Command: A Contribution to Anonymous History* (New York: Oxford University Press, 1948).
150. Ibid., 17–30.
151. Bijker and Law, "General Introduction," 3.

CHAPTER 2

CASE ONE: DOMESTICATING THE DEVIL WAGON: THE INTERCHANGEABILITY OF MUSCLE AND MOTOR POWER

How does a peripheral object become a mainstream device? The first case of this three-part study centers on U.S. automotive history during the first decade of the twentieth century. I attempt to capture some of the dynamics involved in marshalling consumer demand. This study assumes that mass production of automobiles was in progress by 1910. The United States presents an ideal case because of its pervasive car culture.[1]

The automobile at the turn of the century was not only new in form; it was associated with rich outlaws, rogue chauffeurs, and other marginal groups. How did this newfangled machinery of dubious functionality find its way into the lives of a new, rising consumer class? This study focuses on the first decade of the twentieth century with the intent of capturing the automobile's marginal status and the strategies used to overcome it.

Although the widespread dissemination of the automobile may be attributed to the Ford Model T in 1908 and its assembly-line production, this explanation implies that mass production brought about mass consumption. Throughout industrial Europe in the 1920s and 1930s, European manufacturers possessed the same technical capabilities as the Americans, but the success of the automobile in the United States was not replicated in Europe.[2] How was the desire to own cars created in American consumers?

This study makes the assumption, as proposed by John B. Rae, that if mass production is to succeed, mass consumption must already either exist or be imminent.[3] To understand mainstream consumer issues, this study reflects on printed advertisements in consumer magazines. Although printed advertisements reflect the producers' version of consumer interests, they indicate the types of strategies used to transform the newly emerging motorcar into a popular means of transport.

At the turn of the century, advertising began to exert great influence in American life.[4] Ruth Schwartz Cowan, in her study of the mechanization of household implements, argues that advertising lies at the "juncture" between social and technological change.[5] No longer dependent on general merchants to market their products, manufacturers reached consumers directly through advertising, shaping their needs, instilling brand awareness, and opening new avenues for consumer spending.[6]

Advertising allowed publishers to sell magazines at cost or less, which in turn allowed them to dramatically increase their readership and extend their reach into mainstream America. As a result, printed advertisements provide a good source for evidence of the strategies used to target mainstream buyers. In the absence of television and other media, printed advertisements from the first decade of the twentieth century supply a particular perspective on the types of public-relations battles manufacturers had to wage in selling the automobile to the public.

A significant amount of money was spent on direct consumer advertising in the early years despite the relatively small size of the automotive industry: as early as 1907 many millions of dollars were spent on marketing the roughly 40,000 cars sold in that year for prices ranging from $650 to $6,000, with the average selling price at $1,500.[7] In the first six months of 1907 alone, the automobile industry spent $300,000 on advertising in twelve magazines, even though, according to many contemporary estimates, the industry was incurring losses at that time and more than 60 percent of automobile manufacturers failed in the first six years of the twentieth century.[8]

While the lack of advertising metrics for this period makes it difficult to establish causality between marketing strategy and shift in perception, the various elements in printed advertisements certainly document the persuasive measures adopted by early automobile manufacturers. Automobile advertisements in the first decade of the century focused on convincing consumers to buy a car, not to replace an existing one. Thus, the primary task of manufacturers was to show that the automobile was useful and necessary. Competition revolved principally around stealing market share from the horse industry. But how did this shift from muscle to motor power gain momentum?

Pamela Walker Laird, in "The Car Without a Single Weakness: Early Automobile Advertising," indicates that the theme of the automobile's technical superiority dominated early advertisements. She argues that early manufacturers entered the automobile industry because of their "passion for the machines, and for the experiences of automobiling."[9] However, manufacturers refrained from expressing their enthusiasm in their advertising messages.[10] Instead, she argues, the advertisements of early automobile manufacturers featured a "near-universal mechanical theme."[11]

> But apparently automakers did not feel the need to prove that automobility was exciting. All auto ads before 1920, and most before 1930, featured technical discussions appropriate to a new and expensive, exciting but intimidating technology, akin to personal computer advertisements today. Lengthy copy gave potential owners information calculated to inspire confidence in machines.[12]

Laird depicts automotive advertising until the 1920s as concerned primarily with educating customers rather than anticipating and shaping their needs and tastes.[13] Even when automotive brands were associated with prestige, manufacturers de-emphasized the excitement of owning a car.[14] In this sense, Laird argues, automotive advertising, although well funded, was less sophisticated than advertising for other brand-name consumer goods.[15]

Rather than hiring marketing specialists, automotive company owners and chief executive officers wrote their own advertisements well into the 1920s; thus their advertisements "reflected owners' ambitions and concerns to a degree not true for other manufacturers then advertising directly to consumers."[16] This study subscribes to Laird's proposal that early automotive advertisements reflect the thinking of manufacturers.

However, Laird also argues that manufacturers *refrained* from expressing their own passions for automobiling in their advertisements—hence, their staid, technical message. Which personal concerns, then, were expressed in these early automotive advertisements, and which were not? Laird does not address this issue directly but simply alludes to advertisements as reflecting the personal aspirations of manufacturers. This study assumes that these aspirations involved increasing market share. Laird attributes the pervasiveness of the technical, staid tone of automotive advertisements to the carriage industry:

> Instead, the aesthetic roots of auto advertising are to be found in the carriage industry, the automobile's other ancestor, and carriage advertising typically did not picture passengers or try to invoke sensations of speed or motion to appeal to consumers. The carriage trade origins of so many automakers explain the strength of this sedate legacy…Certainly no early auto manufacturers permitted the enthusiasm they expressed about their cars elsewhere to creep into their promotions.[17]

While Laird characterizes these conservative advertisements as appropriate to an exciting new technology,[18] her emphasis on technological excitement seems at odds with the "sedate legacy" of the carriage industry. Scharff, however, describes a slightly different state of affairs: manufacturers/advertisers were keen, rather than reluctant, to adjust their advertising strategies, including their Victorian mind-set, once they realized that features traditionally associated with women also appealed to men.[19]

With this consideration, I used various sources in printed media to verify information gleaned from advertisements. In this chapter I closely examine three major consumer magazines and two trade magazines from the late nineteenth century until 1910. *Collier's Once a Week*, which in 1895 became *Collier's: A National Weekly*, was considered a pioneer for its many articles and advertisements about automobiles in the early twentieth century.[20] *McClure's* (also known as *McClure's Magazine*), on the other hand, although one of the most popular New York–based ten-cent magazines, was conservative and unadventurous in its presentation of materials. Despite its traditional tone, however, *McClure's* featured scientific developments in many of its articles.[21] Hence, these two consumer magazines cover a broad spectrum of attitudes and perceptions.

Life, which began publishing in 1883, provides a check on the claims made by consumer-magazine advertisements. *Life* articulates through illustrations mainstream thoughts and sentiments without the car manufacturer's profit agenda and thus complements written sources found in *McClure's* and *Collier's*. The two trade magazines used, *The Automobile* and *The Horseless Carriage*, were biased in favor of the automobile; for this reason, the kinds of reassurances revealed in their magazines shed light on public concerns about the automobile. They also reported on various activities related to automobiling, particularly in the beginning of the twentieth century. The periodical literature prior to 1910 constitutes the major source of information for the early history of the automobile[22] and presents insights into the transformative elements used to promote the automobile's diffusion.

CREATING AN ANACHRONISM: THE AUTOMOBILE AS A PERIPHERAL OBJECT

"What started the demand for the automobiles, and who first attempted to fill it?"[23] asked Charles E. Duryea, the automotive pioneer who built the first American car in 1893. "Quite a natural question," he continued, "but it is based on a misapprehension. Radical things are never demanded. Improvements are sometimes asked for, but the really great steps in advance are usually so far ahead of the public that they decry rather than ask for them."[24] Duryea described how technological innovations, in his experience, were an uphill battle with the public. As Basalla argues, things come first, followed by the need for them, although this need may take some time to be realized, as Duryea suggested, particularly when inspiring mass consumption.

Contrary to what many historians describe as a ready market poised to replace the horse, prospective users in fact rejected the automobile outright, both as a working mechanism and as a concept. People, Duryea reported, would much "prefer to drive something with life."[25] Thus Duryea took pains to hide his first prototype for fear of being charged a lunatic. Elwood Haynes, who claimed to be the father of the automobile, used a horse to haul his first self-propelling vehicle out into the country-side for a test drive.[26] He thought it unsafe in 1894 to conduct trial runs in the city, as no one had seen anything quite like the vehicle.

Ray Stannard Baker, the first journalist to write about the automobile for a consumer magazine,[27] estimated that there were fewer than thirty self-propelling vehicles in working condition in that year throughout the world.[28] The following year, when Haynes drove through Chicago's Michigan Avenue to the first U.S. automobile racing event, sponsored by the *Chicago Times-Herald*, a policeman ordered him to leave.[29]

In 1900 U.S. census-takers counted 57 automobile factories produc-ing 3,723 automobiles of every kind.[30] The total number of automobiles in existence four years earlier had been only five to six hundred, accord-ing to the notes of a New York appraiser.[31] Some estimated the number to be even smaller, at no more than 200 automobiles in 1898.[32] The U.S. Census Bureau in 1900 found the size of the automotive industry to be so negligible that it did not warrant a separate report.[33]

In highly urbanized areas such as New York City, traffic laws discrimi-nated against this newfangled machinery. A horseless-carriage owner was arrested for driving his vehicle to the entrance of Central Park.[34] Letters to the *Atlantic Monthly*, considered a highbrow magazine, contemptuously described the automobile as a clattering machine, "an anachronism and a blot," disrupting the quiet and bucolic landscape of the countryside.[35] The automobile at this time still required much work to be convincing.

PRICING THE DEVIL WAGON: THE BARRIER TO DIFFUSION?

Some historians, such as Flink and Berger, believe that convincing consumers meant making the automobile affordable. However, high prices do not seem to have posed a significant barrier to adoption. There are indications that automobile manufacturers in fact provided affordable automobiles from the earliest stage of the industry in an effort to lure horse riders. If people earnestly wanted to own an automobile, it was not out of their reach.

As early as 1901 an automobile writer noted that "for those whose purse was short, there were trappy-looking voiturettes."[36] About a decade later, a motorcar company's advertisement, which ran for three years, stated that a car could either be "constructed to humor expensive tastes, or [constructed] to sell on the attractiveness of its price."[37] The nascent automobile industry targeted a wide range of economic classes and sought mainstream users even in its early years.

As early as 1903 a reviewer remarked on the large selection of gasoline automobiles available for sale for anywhere from $500 to $9,000.[38] In 1905 prices of cars with different types of motors ranged from $500 (the Pope-Tribune of Pope Manufacturing Company, Hartford, CT) to $11,000 (the Napier of Napier Motor Company of America, Boston, MA).[39] The price range grew even broader the following year, with prices ranging from $400 (the Orient of Waltham Manufacturing Company, Waltham, MA), a hundred dollars less than the previous year, to thirty times that, at around $12,300 (the Panhard of Panhard & Levassor Automobile Company, France, distributed in New York City).[40]

When we compare car prices with the average income at that time, the automobile industry appears to have tried to appeal to a wide range of buyers. While the U.S. census did not begin to measure annual income until 1940,[41] salary calculations derived from reports compiled from the U.S. Bureau of Economic Analysis show that the ratio of car price to average salary was not far from today's standards. A clerk working in the manufacturing and steam railroad industry from 1905 to 1909 earned an average of $1,076 per year.[42] A federal employee during the same period took home about $1,072 per year.[43]

While these employees may not have been able to pay cash for a brand-new car, car ownership was not out of their reach. An average person intent on owning a car in the early twentieth century would be able to obtain, for instance, a brand-new $500 Pope-Tribune two-seater gasoline car, or even a two-seater $650 Royal electric car, or a four-seater $800 Prescott steam car[44] after a few years of saving or with some form of financing. A department manager selling farm implements, making around $2,000 per year,[45] would be able to afford any one of many different types of cars.

Several years later, options were even less expensive. Throughout 1907 and 1908 the Success Auto-Buggy manufacturing firm frequently advertised in *Collier's* a "patented" four-to-forty mile-per-hour auto-buggy "suitable for city and country use"[46] for $250, about $100 less than the price of a good horse outfit in 1911.[47] The Black Motor Buggy manufacturing firm, also advertising in *Collier's* throughout 1908 and often positioning its advertisement on the front page alongside the table of contents, touted a $375 motor buggy that obtained a gas mileage of thirty miles per gallon[48] on "country roads, hills and mud";[49] their cars were affordable to run as well as to purchase.

At this time, a large number of secondhand cars were also available on the market. An observer in 1905 noted that it was comparatively easy to find a secondhand car of "almost any type at a price very much below its original cost, and in many cases at figures that are really absurdly low."[50] Markets for secondhand models offered cars for as little as $200.[51] Some enthusiasts ventured to build their own automobiles, paying just for parts.[52] Thus the option to purchase a motor vehicle was clearly available to anyone willing to experiment with the newfangled machine. A wide range of prices was available throughout the first decade of the twentieth century for those who sought to own an automobile.

However, public demand did not follow despite the decrease in automobile prices. Flink claims that it was the "inability of the [automotive] industry to produce a low cost vehicle in sufficient quantity [...] that prevented the rapid disappearance of the horse in American cities."[53] However, motorcars that cost less than horses were available well before the Ford Model T. Duryea in fact professed to have designed and created the first U.S. automobile for "people unable to afford horses," something that "ate no oats and caused no expense when not in use,"[54] altruistic sentiments perhaps born of his experience of four years of economic depression following the Great Panic of 1893. However, despite the promises of economic relief and of safer and cleaner streets, the automobile made no practical sense to those for whom it was built.

ANIMATING THE MECHANICAL BEAST: THE AUTOMOBILE ON THE FRINGES OF SOCIETY

What appeared to be an obviously far superior mechanism compared with the horse in the eyes of manufacturers and, for that matter, to historians of the automobile, was certainly not so to its targeted market at this time. For one thing, these strange mechanical beasts appeared hideous. Automobiles were described as "newfangled machinery" that to "yet unaccustomed eyes [was] extremely awkward-looking."[55] Even the fervent automobile advocate C. E. Woods acknowledged—but understated—the problem: "The unsightly appearance of automobiles has been commented on in this country a great deal."[56]

Horseless carriages appeared to the nineteenth-century eye as carriages whose horses had been hacked off. They were ridiculed as having a "carriage-without-a-horse look" or as "shaftless bugg[ies]."[57] Their mutilated appearance provided such a spectacle that the very first cars produced in the United States toured with a circus alongside elephants and trapeze performers (Figure 3).

The advertisement for the Haynes gasoline car claimed that the company's pioneer models had been "star attractions at country fairs and

FIGURE 3. Motorcars as circus exhibits.

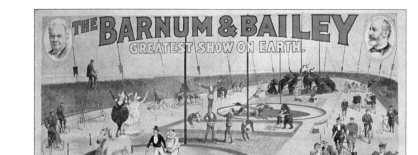

Source. Image courtesy of Circus World Museum, Baraboo, Wisconsin, with permission from *Ringling Bros. and Barnum & Bailey®.*

drew huge crowds."[58] Horseless carriages appeared at the 1900 Paris Exposition, a venue famous for the display of creations conceived beyond the "wildest flights of imagination."[59] The novelty of the horse-less carriage was such that people saw it as a curiosity rather than as a practical, usable tool.

A writer recalled that "an automobile drawn up at a city curb was a sight to attract a gaping crowd."[60] One observer sympathetic to the cause of automobiling even admitted that "many models shown were not quite acceptable to the eye. Perhaps for a lack of previous education we are still feeling the want of a horse in front of some of these odd-looking traps."[61] In an effort to make the motorcar look more familiar, an inven-tor in 1900 attached to the dashboard an imitation horse's head, which he claimed doubled as a gasoline tank.[62] The same general idea persisted five years later in a *Life* illustration of a motor vehicle posing as a horse and buggy (Figure 1).[63] The automobile stood out as an irregular piece of machinery because of its lack of a horse.

The dramatic effect of the early horseless carriage was also heightened by the various sights, sounds, and smells associated with it. Horseless carriages running on steam often left a trail of vapor and, in some unfortunate cases, thunderous explosions from malfunctioning boilers. A competitor in the famous Paris–Marseilles–Paris race in 1896 described how his steam engine "need[ed] a mechanic as fireman" in addition to one in charge of automotive maintenance.[64] His horseless carriage loomed so large it frightened two drivers off the road, causing them to overturn their own carts.[65] With the more modest-sized steam engine, the driver typically sat on top of his boiler, which contained about 400 pounds of pressured steam; it was, according to one observer, "a toss up whether he was going to travel straight ahead or straight up."[66]

Petroleum motorcars, on the other hand, convulsed like beasts, threatening fire from their fuel. A Cornell University professor in 1901 assured users in all seriousness that petroleum motorcars "may be handled by the amateur with no other special danger than that of accident resulting in the firing of his tank."[67] Even the esteemed first-class Daimler, renowned for its reliability, used a tube ignition[68] in 1900 that easily caught fire, endangering both the car and its occupants.[69] Accidents and mechanical malfunctions were the norm. An 1898 cartoon depicting a horseless carriage as a dragon attacking a man[70] speaks of the way early adopters struggled with their machines.

In addition to mechanical hazards, other annoyances plagued early automobiling. Motor carriages running on petroleum reeked of foul odors from the wastes of oil combustion. A 1903 *Life* illustration depicted a man strapped behind his car and told of his finally having the chance to "enjoy the sweet odors emitted" by his own machine.[71] Such illustrations mocked the pervasive stench arising from automotive motors. Cars at this time also vibrated vigorously, an observer noted in 1896, from the rapid explosions in their cylinders.[72] Startling noises rose from carriages running on compressed air.[73] Early motor engines made so much noise that an observer noted the futility of having a horn; in fact, early models did not have one.[74]

Electric cars running for more than twenty-five miles threatened to spill acid from their accumulators.[75] Although they tended to be less noisy, less odorous, and less jarring compared with petroleum vehicles, they were heavier because of their storage batteries and were more expensive, limited in range, and impractical outside cities, where current was not readily available. Petroleum motorcars performed well at all distances and speeds, but their odors were highly unpleasant, and they were difficult to start, terribly noisy, and rough in motion. Steam engines used cheaper and more readily available fuel, but they required a fireman and a mechanic to keep their boilers from exploding because of too much pressure. They also corroded easily, with sediments building up frequently.

All these unpleasant properties arose from the means necessary to animate the monstrous machines. These disconcerting sights, sounds, smells, and convulsions constituted, according to a pro-car sympathizer, the main objections to motorized carriages.[76] The notion of an animated machine designed to become part of daily life must have been immensely disturbing. Schivelbusch writes of a similar experience in England in the early years of railways:

> The popular images of the "mechanical horse" manifest fear in the very act of seeming to bury it in a domesticating metaphor: fear of displacement of familiar nature by a fire-snorting machine with its own internal source of power.[77]

The average American citizen of this period viewed the mechanical horse invading the streets with similar fearsome apprehension.

DEALING WITH DEATH MASKS AND DEMONS

It did not help matters that early automobile adopters appeared menacing behind the wheel. Even the most ardent pro-car magazine called the automobile fashions of the day "hideous."[78] This apparel, said to be in demand in many high-fashion cities such as Paris and New York, included a face mask covering the neck, with holes for the eyes fitted with goggles (Figure 4).

An observer commented that the most beautiful woman wearing this face mask would have resembled, in his words, "a three-ring circus or Mardi Gras fête."[79] The gowns accompanying these masks covered the entire body, rendering the drivers unrecognizable. Automobile gowns typically were made from wool or other heavy fabric, with rubber and heavy leather styling, a popular choice at that time (Figure 5).[80]

Even if a pro-car writer wished to be complimentary, the fiendish effect of such outfits could not be overlooked. An article meant to pay homage to the race-car driver Henri Fournier, winner of the 1901 Paris-to-Berlin race, described him as "uncanny" in his big black goggles, which rendered his "outward appearance [that of] some new sort of demon."[81] The ungainly costumes meant to protect motorists from mud and dust created such an ominous effect that even the premier pro-car magazine of that time could not resist publishing a lampoon that depicted a couple in their automobile outfits inducing great fright in their own child.[82] The headgear meant to protect the face from the

FIGURE 4. Driver's headgear of 1907.

Source. The Automobile, September 19, 1907, 403.

FIGURE 5. Fashionable automobile outfits of 1905.

Source. The Automobile, January 14, 1905, 35.
Note. Captions read (l) "Jacket and adjustable skirt of dark material, red waistcoat with gold buttons." (r) "Blue woolen driving suit 'Micaud', considered very 'up-to-date', edged with green leather. Note helmet-shaped hood."

elements became known as a "death mask." And, indeed, such masks lived up to their name (Figure 6).

The "death mask" rendered drivers unidentifiable and hence unaccountable for any fatalities caused by their driving. Goggles with face masks became so popular that "all motorists with a love for fast driving came to look uniformly alike," noted a writer on traffic laws and violations.[83] A joke published in *Life* in 1901 made the point:

> Automobilist: Say, I want this mask changed. It doesn't cover my face enough.
> Clerk: But it's the regular thing.
> [Automobilist:] Can't help that. I find that the people I run over are apt to recognize me.[84]

FIGURE 6. Death masks.

Source. The Automobile, July 11, 1903, 29.
Note. Captions read (A) "Costumes offered in New York." (B) "Death mask and veil head gear."

Another derisive but nonetheless apt illustration of this notoriety depicted a mother bidding farewell to her daughter with the usual motherly reminders, one of which was to flee the instant she happened to run over a child to avoid getting her name in the papers (Figure 7).[85]

Another equally suggestive joke, published the following year, made the same point. "He is the champion of our automobile club," says a driver as a haughty-looking man drives by in the opposite lane. "Yes?" the passenger urges the driver to continue. "Yes," says the driver, "he has killed more people without getting his name in the papers than any other member."[86] Indeed, the bloodthirsty image of the early automobile driver came to be intimately associated with the concept of the horseless carriage itself. Many illustrations and commentaries characterized automobiling as being as diabolical as its masked drivers (Figure 8).

The gleeful face of the driver certainly emphasized amusement at the expense of public safety and perhaps created a need for the sedate tone of automotive advertisements as described by Laird. In the interest of winning

FIGURE 7. *Life* illustration.

Source. Life, November 21, 1901, 415.
Note. Caption reads "Now, daughter dear, if you should happen to run over any children, do hurry away before they get your name in the paper!"

FIGURE 8. *Life* illustration.

Source. Life, January 23, 1902, 63.

mainstream users, it may be that manufacturers wished to avoid further antagonizing the public and instead sought to recast the automobile's villainous image with the staid aura of the horse-drawn carriages. It did not help the cause of the automobile that many affluent drivers perhaps scorned the plight of the average pedestrian (Figure 9). Reckless rich joyriders pushed the automobile even farther from the mainstream (Figure 10).

A 1903 Dewar's Scotch advertisement best captures heedless automobiling in its early days with its promise that "there is no more exhilarating sport or recreation than automobiling. The pleasure of a spin over country roads or through a city park is greatly enhanced if the basket is well stocked with Dewar's Scotch"[87]—again, imagery in sharp contrast to the staid mechanical theme of automobile advertisements described by Laird. Automotive manufacturers evidently did not need to portray the excitement of automobiling; other businesses did that for them (Figure 11).

FIGURE 9. *Life* illustration.

Source. Life, July 3, 1902, 9.
Note. Caption reads "The quick or the dead!"

FIGURE 10. *Life* illustration.

Source. *Life*, October 9, 1902, 303.

The emphasis on the exhilaration of drinking and driving is indicative of the reckless spirit of early adopters who used the automobile for amusement. Some examples of the great disconnect between the automobile

FIGURE **11.** Dewar's advertisement.

Source. Life, July 9, 1903, 34 (inside front cover).

and the public were captured in the many illustrations of the rich amusing themselves at the expense of common folk (Figures 12 and 13).

Titles of illustrations lampooning the rich included "Joy-riders drive on after running down boy." A *Life* cartoon suggested physically isolating automobiles within the close confines of a horse race arena, dubbed "Speedway for Millionaires Only."[88] The sense of entitlement of the rich and their wanton disregard for the welfare of average citizens inspired anger and public outrage. Automobiles came to be associated with debauchery and delinquency.

Hatred for the automobile gained such fervor that it became politically astute to condemn these devil wagons. For instance, in New York City, policemen reportedly arrested automobilists indiscriminately in order

FIGURE **12.** *Life* illustration.

Source. Life, November 20, 1902, 439.
Note. Caption reads "THE PROPER SPIRIT. 'Arrh! Get off the earth!' 'I am, sir. May I come down after you have passed?'"

FIGURE 13. *Life* illustration.

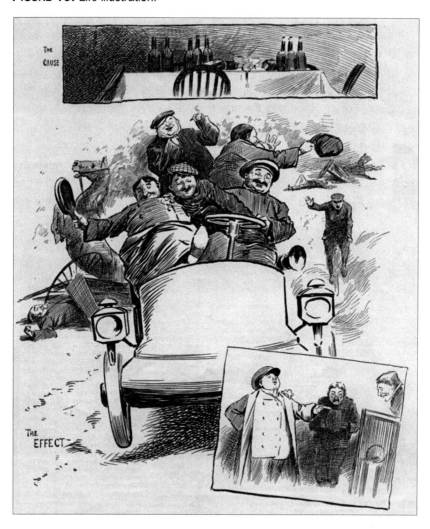

Source. Life, October 2, 1902, 279.
Note. Caption reads "THE CAUSE. THE EFFECT. MONEY TALKS. On being arrested, he presents his card, and the sergeant and policeman are allowed to apologize."

to please the "higher up" who, playing politics, sought to please the majority of voters who had been "down on" the automobile.[89] Thus, despite the promise of cleaner streets and other public health benefits, governmental officials were adamantly against the automobile and used governmental resources to curb its use.

CAPTURING THE SLAUGHTERER

The media were equally unsympathetic. Social commentaries in the forms of illustration, satire, and poetry evoked images of the automobile as a "slaughterer" that tossed common pedestrians, including children, into the air (Figures 14, 15, and 16).

Parodies of poems such as Edgar Allan Poe's "Valley of the Unrest" and Alfred Tennyson's "Charge of the Light Brigade" depicted trails of death and damage left by the automobile.[90] The personal automobile of the millionaire William K. Vanderbilt was named the "White Ghost."[91] In his review of the famous Vanderbilt Cup race of 1908, a journalist began with a satirical note of astonishment that only one child was bruised and broken. He was expecting hundreds.[92]

Cars posed hazards even for their drivers. An observer of the Vanderbilt Cup pointed out that "there was not a second while the race was on that any driver of any of the eighteen cars might not for a dozen different causes have been hurled into eternity."[93] In one of the earliest studies of automobile accidents, survey results showed that 202 people were injured or killed in 96 accidents during the three summer months of 1907 alone—a number that a pro-car sympathizer deemed "really astonishingly small" given the amount of traffic at that time.[94] Before this formal study on automotive accidents, social commentaries had already alerted the public to what was happening in the streets.

A score of automobilists, many of them prominent businessmen and socialites, became fugitives from the police.[95] These costumed drivers came to be associated with outlaws, as aptly illustrated in a 1903 cartoon (Figure 17).

The public, indignant about reckless driving, condemned the automobile as a killing machine. One notorious case was that of Frank Hodge, who in 1905 ran down a man in Pittsburgh in a manner witnesses described as "bold and cold blooded."[96] Hodge fled on the night of the accident to Buffalo, where his family resided. After the Pittsburgh police tracked him down on charges of felonious assault and battery, he surrendered himself to the Buffalo police but was immediately released on a mere $1,000 bail after some maneuvering by his prominent family. Pittsburgh residents became so enraged over what had been a long series of cold-blooded automobile killings in their city that steps were taken to extradite the young man to Pittsburgh for trial.

FIGURE 14. *Life* illustration.

Source. *Life*, December 4, 1902, 490.
Note. Caption reads "A NEW GAME. This winter's new indoor game, devised to take the place of the summer automobile, as ping-pong did that of lawn tennis. Played with live children, instead of dolls, the game is simply irresistible."

FIGURE 15. *Life* illustration.

Source. Life, August 21, 1902, 151.
Note. Caption reads "This scheme of Cholly Billions admits of a high rate of speed, while reducing to a minimum all risks of damage suits."

FIGURE 16. *Life* illustration.

Source. Life, December 19, 1901, 533.
Note. Caption reads "Why interrupt the gentleman when he is having his fun? A fixed rate for the loss of a limb could be settled without the annoyance of having to stop the machine."

FIGURE **17.** *Life* illustration.

Source. Life, July 30, 1903, 113.
Note. Caption reads "HE READ THE NEWSPAPERS. *Western Badman (genially)*: Shake, Pardner! We ain't entirely outgrown guns in this section, an' we don't wear our masks every day, but we can appreciate progressive methods. How many notches ye got in yer machine?"

Even the famous race-car driver Henri Fournier was arrested for fast driving in New York City while demonstrating a car to a prospective buyer.[97] That same night, a chauffeur was arrested after being chased around New York City, where he allegedly intentionally knocked down a policeman on a bicycle.[98] The chauffeur was charged with felonious assault and fast driving, with bail set at $2,000. When his employer went to the courthouse to bail him out, he learned of his chauffeur's behavior and left in disgust without making the necessary deposit. Chauffeurs constituted a group of users who shared reckless driving behavior with the rich but had opportunistic motivations as well. They took advantage of their access to large cars to go joyriding, causing accidents and embarrassing their employers with their outright defiance. The motorcar came to be associated with physical as well as social disorder.

PUNISHING RECALCITRANT CHAUFFEURS

Kevin Borg describes the problems wealthy employers faced as a result of their dependence on recalcitrant chauffeurs:

> Chauffeurs became a serious problem for wealthy motorists during the first decade of the twentieth century. They extorted commissions and kickbacks from garage owners, took their employers' cars out for joyrides at all hours, and exhibited a brazen disregard for social decorum. They did not behave as servants.[99]

Chauffeurs were hired to drive and maintain large gasoline and steam-powered cars on the assumption that they would behave like coachmen. Borg describes the attempt to transfer social practices from horses to horseless carriages in a social structure where the balance of power had changed:

> This is exactly what many wealthy motorists did when they purchased an automobile: they hired chauffeurs, dressed them in livery, and gave them responsibility for the care and maintenance of their vehicles, transposing the rules associated with horse transportation to their new horseless carriages.[100]

In the case of the recalcitrant chauffeurs, the displacement of the coachman was far from straightforward. The mechanical problems early motorcars presented, Borg argues, forced wealthy motorists to depend on their chauffeurs.[101] The scarcity of qualified chauffeurs and the specialized knowledge required to maintain a motorcar allowed chauffeurs to disrupt the master-servant hierarchy. A chauffeur's wage in 1906, anywhere from $75 to $150 per month,[102] would have easily paid for a $250 Success Auto-Buggy in two months' time. Although well compensated, chauffeurs preferred to drive the larger and more powerful cars of their employers without obtaining consent.

According to a *Life* editorial, "the chauffeur owns the highway. Out of danger himself, drunk with speed, absolutely irresponsible, and always fleeter than his outraged victims, he devastates the country, and in perfect safety."[103] An attorney for a large city garage observed that many chauffeurs who were discharged for drunkenness, theft, or reckless driving found new employment again and again because employers did not bother with background checks.[104]

One of the ways wealthy motorists eventually controlled the chauffeur problem was through legislation. Wealthy Pennsylvania employers, for instance, were freed from any responsibility for damage done by their autos if they assisted in the prosecution of their joyriding chauffeurs.[105] Many states began imposing age requirements (18 years) and the testing and licensing of all chauffeurs.[106] Wealthy motorists also addressed their dependence on their chauffeurs' technical skills by opening YMCA training schools that generated alternative sources of skilled labor.[107] Garage owners began dealing with car owners directly, instituting stricter accounting procedures, and closely monitoring activities in their garages.[108]

Thus wealthy car owners, although seeking to promote the use of automobiles, did not share the interests and values of other users such as reckless rich joyriders and recalcitrant chauffeurs. Judges who favored automobiles sentenced chauffeurs to five days in jail and a $50 fine for speeding violations.[109] In one famous 1905 case, a judge professing to be an ardent automobilist sentenced a chauffeur to eighteen months in prison (the full penalty of the law was two years in jail) plus a $100 fine for accidentally running over a little boy.[110] Despite a recommendation of mercy by the jury and hopes of an acquittal, the severity of the punishment, according to the judge, served to warn and admonish automobile drivers.[111]

In another case four years later, a police judge in the Indianapolis city court penalized a chauffeur with sixty days in prison and a $1 fine for driving while intoxicated; another $1 for profanity; $200 for malicious trespass for operating an automobile without permission, and an additional sixty days in prison; an additional $200 plus another sixty days in prison (180 days altogether) for malicious trespass for the damage done to the cab; and a $50 fine for violating speed laws.[112] Although some judges themselves owned cars, they wished to send a clear message to reckless drivers. Chauffeurs were to be severely punished for reckless driving, with the aid of the very same owners who favored fast cars.

The case of the recalcitrant chauffeurs shows that in displacing the culture of the coachman, wealthy motorists failed to transfer practices seamlessly in an unstable social structure. It was no longer just a matter of motor power replacing muscle power within a similar set of circumstances. The master-servant relationship became uncertain as the technically able chauffeur sought to challenge entrenched work and social relations established under a coach-driven society. Later discussions will show that co-optation rather than confrontation facilitated the success of motorization. In the case of the recalcitrant chauffeurs, what they had definitively accomplished was to further antagonize the public and stigmatize the devil wagon with their hazardous driving.

ARTICULATING AMBIGUITY AMONG AUTOMOBILE ADVOCATES

Many early users of automobiles attained such notoriety that mainstream society became alienated from automobiling. A physician established an insane asylum with a separate ward just for deranged motorists and chauffeurs.[113] The two social groups publicly associated with the automobile, the reckless rich who amused themselves at the expense of the average citizen and the recalcitrant chauffeurs who followed suit at the expense of their wealthy employers, shared the meaning of the automobile as a source of pleasure but ultimately were motivated differently.

The rich driver used the automobile simply as a diversion, whereas the chauffeur used it to escape from his subservient role. This complexity in grouping high-speed drivers according to shared meanings poses a challenge in the application of the SCOT model. Should shared passions for high performance cars constitute one social group, or should the master and servant categories be separated into two groups according to their motivations? The organizational task would be even more daunting if reckless horse riders were also assigned to the first, single social group, as reckless drivers who shared a love for speed, as *The Automobile* noted.

The Automobile, known for its pro-motorcar bias, sought to shift the blame from the automobile to its operator, arguing that reckless drivers were previously reckless horsemen: "Practically all those who now own machines were horse owners before, and if they are discourteous now, they were when they drove a horse."[114] Thus, the implicit message was that the motorcar itself did not cause bad behavior; reckless horse riders did.

The owner of the *Chicago Times-Herald*, who sponsored the first car race in the United States in 1895, published statistics characterizing horses as unreliable and dangerous. The study showed that horses were the leading cause of fatalities and accidents (40%) compared with the meager 5% caused by automobiles.[115] The study, however, did not indicate the percentage of automobiles in active use that had been involved in an accident; the number of automobiles was a small fraction of the millions of horses used at this time.

Nonetheless, the pro-car article went on to describe the horse as aggressive and dangerous, as well as cowardly and stupid, because it would "take fright at a fluttering bit of paper or some equally harmless thing and run away."[116] This attack on the horse came as an offensive move to address the widely held belief that automobiles were monstrous killing machines. In these early days, if one favored automobiles, then one must be against horses.

UNVEILING THE UNTHINKING MACHINES

Horses had long been perceived as a "friend of man" in work and leisure.[117] Automobiles, on the other hand, frightened horses and pedestrians; many people were maimed or killed because of reckless driving and malfunctioning motors. Thus, traffic laws were written to protect horses from motors. A Massachusetts automobile law in 1902, for instance, required automobile drivers to stop at the behest of drivers of horses if the latter found their animals frightened.[118] Other states such as New York, Massachusetts, and Michigan required automotive speed to be no more than ten miles per hour within city limits.[119]

Articles and illustrations in those early days depicted automobiles as highly unreliable, dangerous, and menacing to society—in essence, unlike the horse. An 1896 writer contrasted sentient horses with unthinking machines; he predicted that people failing to appreciate the life in horses would readily adopt the automobile.

> To tell the truth, all mankind may, with great clearness, be divided into two parts—those who understand horses and those who do not. These are people who will drive or ride a nag all day, nay, who may own one and use it for years, whose powers of observation are not sufficiently enlisted in the details of the animal to distinguish it from any strange horse in the next stall, unless there be some gross difference in color. Such equestrians will be content to see a fine horse, with nerves, eyes, muscles, and possibilities for good or evil, cashiered in favor of the dead certainty of a peripatetic steam-engine.[120]

Automobiles were lifeless machines, according to the writer, whose only place would be to perform tasks too inhumane for horses:

> But is it not absurd to defend a good horse from a horseless carriage? Each will have its appointed duties, and no one will be so glad as the man that makes a friend of his nag that a nerveless substitute has been found for the straining, scrambling, jaded creatures which afford such heart-breaking scenes on the icy cobble-stones of the city.[121]

This horse advocate suggested a compromise, although he recommended a far inferior status for the "nerveless," unthinking machine. Advocates for the horse believed that losing the horse would mean dispensing with certain work routines. A historian in 1897 proclaimed, "The substitution of inanimate power for the animal power on which our race was formerly dependent means a separation of the force which does the work from the intellect which directs it,"[122] an argument that resembles the general criticism on automating human labor. People wanted to drive something with life, as Duryea himself observed.[123] The lack of inherent intelligence in machines meant that the driver would have to work harder.

An observer stated, "The man who drives a horse has little to do; the horse finds the way and does the work, but the driver of a motor carriage has a senseless machine, and all direction must come from him."[124] Some technological analysts, such as Latour, consider machines more reliable than human workers, particularly in performing repetitive tasks. [125] Many people in the nineteenth century disagreed. The arrival of motorized machines required humans to be even more alert and skilled because they could no longer rely on the horse to compensate for their lack of concentration. The experience of replacing the horse meant losing some advantages, such as its abilities to find its way home and to avoid accidents without the need for constant direction from the driver.

Racing commentaries on the first U.S. car race in 1895 agreed that horseless carriages would be more prone to accidents because they lack the navigational instinct of experienced horses. A reporter of the race articulated their line of thought:

> It should be borne in mind that the carriage without a horse is also without the convenience of a horse's intelligence, which really in ordinary traffic and driving relieves the person holding the reins of a large part of his responsibility. For not even the best made motor can think, and the slightest carelessness on the part of its driver, or failure of the guiding apparatus, might precipitate an accident.[126]

The driver of a motorized car would be alone, without aid; there was no horse to compensate for him. Accidents then became associated with machines, which by virtue of their lack of instinct were unable to protect and assist drivers.

The predominant dynamics in early automobiling concerned the credibility of the automobile itself, as both a concept and a device, and its ability to deliver the benefits the horse provided. The predominant question in people's minds was "Why do we need the automobile when we have the horse?" Moving the automobile from periphery to mainstream hence required doing away with the horse. C. E. Woods,[127] an avid machinist and most likely the same C. E. Woods mentioned in the cover story of *McClure's Magazine* and described as one of the leading automobile manufacturers in 1899,[128] insisted that the work of the horse had to be directly replaced by a contrivance able to perform the same tasks. This manufacturer firmly believed that motor power would soon displace muscle power.

> When we review all that has been done by mechanical devices toward the displacement of animal power, it is very hard to refrain from drawing a conclusion that the horse must go; that is, speaking in the broad sense of the word. Mechanically propelled vehicles for all purposes are here.[129]

However, the horse industry remained strong and lucrative despite the arrival of the motorcar. In fact, the demand for horses increased.

INVADING THE HORSE INDUSTRY

The annual production of horse-drawn carriages had been substantial. In 1900 a total of 907,482 family-use carriages alone were produced, worth $51.5 million.[130] Added to this number were 575,351 business wagons, with another 2,316 for public transportation, which together amounted to roughly $32.6 million in sales,[131] for a total of 1.5 million horse-drawn carriages of all types, valued at $84 million.

Horses at this time fueled much of the rural and urban economy. U.S. Department of Agriculture statistics place the number of horses at roughly 14.2 million in 1890, and at over 17 million in 1905, with their selling price almost doubling, from $37.50 per horse in 1899 to $70.34 in 1905, during the supposed period of the "passing of the horse."[132] When the automobile began to be marketed to the public, there was a simultaneous increase in the demand for horses. According to Clay McShane and Joel Tarr,

> Between 1870 and 1900, as society became more dependent on the horse, the amount of capital invested and the number of workers employed in industries such as carriage-making and repair, saddlery and harnesses, and whip manufacture vastly increased.[133]

It is possible that the burgeoning horse industry whetted the appetite of automobile manufacturers, who sought to tap into this lucrative market. The autumn horse show of 1896 has been described as having increased in importance, one year after the first automobile race in the United States.[134] Even as late as 1908 the demand for horses steadily increased, with numbers reaching close to 20 million.[135] With an eye toward this lucrative market, automobile manufacturers did not intend for automobiles to perform types of work different from those performed by horses; they wanted to plunder the lucrative horse industry. A writer articulated this desire:

> Naturally there will always remain a limited number of users who, by preference or supposed economy, will remain true to the horse and buggy, but the total given [937,000 family and pleasure carriages], which represents a value of $55,000,000, shows that *there is still a tremendous field which the automobile builder may invade and reasonably hope to capture* by quicker, cheaper and more economical methods of mechanical transport (emphasis mine).[136]

Stealing market share from the horse was not the only motivating factor. Manufacturers wanted the automobile to be perceived as a necessity, rather than a plaything, to ensure its perpetual use. A letter to the editor of *The Automobile* astutely articulated the thoughts of many manufacturers: "When the faddist has worked out his pleasure car, he may or may not replace it; when the motorcar is used in business, it *must* be replaced by another."[137] Automobile manufacturers had much work to do to change mainstream perceptions of their infamous product.

MOVING FROM THE PERIPHERY TO THE CENTER OF SOCIETY

C. E. Woods, who published a manual on the construction, care, and operation of the electric automobile in 1900, asked the same question Duryea had:

> What conditions exist that will make a market for automobiles or create a desire in the public mind for their use? Some will say progression, the spirit of which surrounds us everywhere; others say, expediency and the desire for saving minutes and even seconds; others, again, their convenience and readiness for instant use; all of which are true but do not in a broad sense answer the question, but create another as to what has made all these things desirable on the part of the public as things necessary to its comfort and welfare.[138]

Comfort superseding other, more exciting features such as "expediency" and "readiness for instant use" would require the automobile, considered a fad rather than a necessity, to shed its whimsical image in order to become a staid, reliable form of transport. One way to accomplish this transformation was to associate the devil wagon with the most accepted form of motive power—the horse. Thus at this early stage manufacturers had to emphasize the automobile's similarity to the horse rather than its novelty. At the same time, they also sought to displace the horse by highlighting its biological limitations.

Motorized power posing as muscle power came as a move to pull the automobile from the periphery to the very center of society—automobile manufacturers employed various means to project the image of a reliable, easy-to-operate, mundane, practical machine rather than that of a foul mechanical monstrosity that compromised public safety. As early as 1900 manufacturers such as Woods were already articulating their desire to see automobiles provide services identical to those of horse-drawn carriages without any sacrifice in ride quality:

> There is a mistaken idea with many people, who have not given the subject any thought, that automobiles are sold for their novelty and because they go without a horse. But this is wrong. The purchasing public which uses automobiles buys them primarily

> for the same purpose for which it has always purchased any class of vehicle, namely, because a carriage or a vehicle is needed for personal transportation, convenience and comfort; and as it is among the better class of carriage users that automobiles are generally sold, they demand the same diversity of design, the same elegance in finish, the same magnificence in appointment, and the same easy riding qualities that they have always been accustomed to when drawn by horses.[139]

The pleasure of comfortable riding and the practical necessity of mobility embodied what manufacturers sought to display in their automobiles. In early automobile construction, rider comfort received minimum consideration until carriage mechanics and builders stepped in to help popularize the automobile by creating larger and roomier seats, adjusting automotive springs for even weight distribution, and installing side doors for easy access.[140]

The author of a letter to the editor of *The Automobile* observed that the move to hire carriage mechanics came from the need to decrease the weight of automotive parts in order to make them look less like heavy machinery and to allow them to have "a pleasing effect to the eye" through adjustments to the dimensions of the body for symmetry.[141] The president of the Carriage Builders National Association admitted in his opening day speech at the organization's 1905 convention that "the carriage maker has already been called on by the engine builder to equip his machinery with durable and luxurious bodies and upholstering, and those who have taken up this new branch of industry report that it is constantly increasing."[142] Thus the automobile had to undergo a radical shift from the noisy, the dangerous, and the curious to the respectable, the comfortable, and the sensible personal transport of the everyday.

One could argue that this transformation involved making the horseless carriage "invisible,"[143] and hence, in Roger Silverstone's terminology, "domesticated," by transforming a radically new device into something ordinary, taken-for-granted, and part of everyday life. One way this invisibility was accomplished was through the borrowing of elements from the horse-drawn carriage industry. Traditional names and designs from the carriage industry were copied by the automobile industry. The phaeton, a well-known horse-drawn carriage with four seats, designed for open air with a portable half-top or without a top, was copied by many automobile manufacturers in their standard models.[144] Other examples include runabout automobiles, which came from runabout horse-drawn carriages. The coach, a generic term used to describe a closed vehicle that could seat four or more people,[145] particularly dominated the later versions of the automobile as manufacturers sought to emphasize comfort and privacy.

The effort to duplicate horse-drawn carriages in almost all respects speaks of the conceptual dependence of the nascent automotive technology on the object it sought to displace. A 1905 manual for the construction of practical and workable light motor carriages recommended using the body frames from horse-drawn carriages.[146] *The Automobile*, a staunch proponent of motorized vehicles, admitted in 1906 that "without endeavoring to improve on the fundamental principles of carriage construction," horseless carriages could be built by merely "applying mechanical power for propulsion" to horse-drawn carriages.[147] *The Automobile* described the survival of horseless carriages as a matter of their mimicking rather than differentiating themselves from their competitors. Car designs were translated into an equine idiom: the bodies of early motorcars were purposely constructed to resemble horse-drawn carriages[148] (Figures 18 through 23).

Conjuring Interchangeability in Form at a Cheaper Price

The Automobile in 1903 observed that carriage makers would be forced to start manufacturing automobile bodies or risk losing their best artists and workmen to the automotive business. Such dilemmas were indicative of the motorcar's invasion of the socioeconomic infrastructure built around the horse. The carriage maker had to choose between selling his business to an automotive manufacturer and joining the industry himself.

> *In practice he generally chooses to become financially interested in an automobile firm or to sell out to one.* In either case there is a merging of carriage traditions and new automobile requirements, which will result in more attention being paid to distinctive styles in automobiles and appropriate names for each of them. And nothing is admittedly more difficult than to devise brand new names for new commodities, *those developed in the carriage industry will be preserved in so far as possible* (emphasis mine).[149]

The writer described the persistence of the carriage industry's brand names and styles through the purposeful copying of the automobile industry. However, as with horse-drawn carriages, there were no uniformly accepted definitions of styles and classes in motorcars.[150] For example, in the case of the runabout—among the simpler carriages designed to carry two passengers—even professionals found it difficult to agree on its standard features.[151]

Despite the collaboration of several hundreds of men to standardize carriages, they "apparently [were] not [...] able to discover a uniformly accepted basis for definitions of the various styles or classes of carriages"[152] (Figures 24 and 25). Runabouts were generally described as "light, handy open wagons," but "chelsea" cars also came to assume the same characteristics, creating much overlap.[153] Despite the lack of uniformity and some confusion, the automobile manufacturers persisted in copying the carriage industry's product scheme, making new things old.

A resemblance between the motorcar and the horse, including the horse-drawn carriage, had been made as a way to domesticate the automobile into everyday life. While Silverstone speaks largely of individual and social group experience, such as families imprinting their own mark on new technological artifacts,[154] this chapter speaks more of methods used to change the general public's impressions of the automobile and the efforts exerted to overcome its notorious reputation.

Early adopters of the automobile were in precarious company. On the one hand, a recalcitrant but apathetic group of users, such as the reckless rich and irresponsible chauffeurs, generated bad publicity for the automobile; on the other hand, a passionate group of users, such as

FIGURE **18.** Simplest carriages.

Sources. (A): *Life*, February 21, 1907, 260 (from advertisement for Kelly-Springfield Tires). (B) *The Automobile*, April 18, 1903, 422.

FIGURE **19.** Buggies.

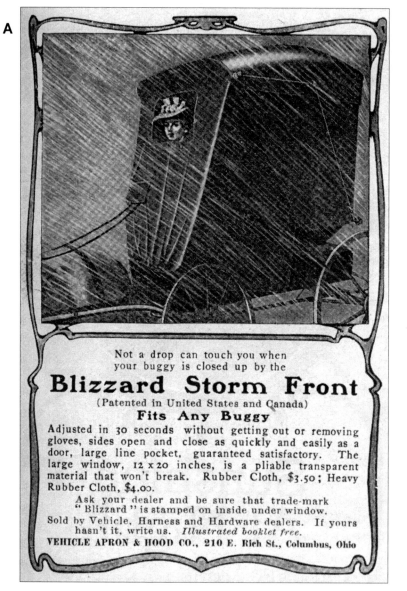

A

(continued on next page)

FIGURE **19.** (*continued*)

Sources. (A) *McClure's*, April 1904, 115. (B) *Life*, May 16, 1907, 670.

FIGURE 20. Surreys.

A

No. 331. Canopy Top Surrey. Price complete, $65.50. As good as sells for $25 more.

33 YEARS
SELLING DIRECT

Our vehicles and harness have been sold direct from our factory to user for a third of a century. We ship for examination, approval, and guarantee safe delivery. You are out nothing if not satisfied as to style, quality and price. **We are the largest manufacturers in the world** selling to the consumer exclusively. We make 200 styles of Vehicles, 65 styles of Harness. Send for large free catalog.

Elkhart Carriage and Harness Mfg. Co.
Elkhart, Indiana

No. 10. Single collar and Hame Harness. Price complete, $14.60. As good as sells for $5. to $8. more.

(continued on next page)

FIGURE 20. (*continued*)

Sources. (A) *Ladies Home Journal*, March 1906, 48. (B) *Life*, January 19, 1905, 62.

FIGURE **21.** Stanhopes.

A | **Spider Stanhope**
Seashore or Country

Resorters, golfers, town or village residents find the "Spider Stanhope" the fittest for their needs. Perfectly correct, it combines in harmony the style and convenience of the Stanhope with the Spider characteristics—Dickey seat— new style dash—rich and graceful scrolled body loops resting on rubber—head elliptical springs.

Our free booklet, "That Carriage Matter," saves you money by buying direct, tells the carriage buyer and illustrates beautifully vehicles of proper style for every need in city or country, and gives reliable and valuable carriage information. Send for it—postal brings it.

MORRIS WOODHULL, Mfr., Home Ave. and 5th St., Dayton, O.

(continued on next page)

the owner of the *Chicago Times-Herald*, sought to generate goodwill for the automobile but attacked everyone else—at least, anyone who caused adverse publicity for the automobile.

FIGURE 21. (continued)

Sources. (A) *The Saturday Evening Post*, May 6, 1899, 717. (B) *Life*, May 16, 1907, 671.

FIGURE 22. Phaetons.

(continued on next page)

The unpleasant effects of bad publicity were such that any buyer in 1903, for instance, was assured by articles such as those published in *The Automobile* of the possibility of purchasing an automobile without attracting too much attention.

> The purchaser can, at the same price [$500], also secure an automobile carriage in its simplest form, fitted with piano-box body and upholstered seat with capacity for two persons sitting side by side. *This has the general appearance of the light horse-drawn road wagon,* and, complete with its power plant, weighs but 550 pounds (emphasis mine).[155]

It is also possible that the copying of the designs of horse-drawn wagons emerged seamlessly from preexisting practices of the horse industry. Indeed, for those brave souls who sought to try the new mechanical device, the art of purchasing automobiles was patterned on the art of purchasing horse-drawn carriages. Laird describes early advertisements involving

FIGURE 22. (*continued*)

Sources. (A) Reprinted with permission from Heart Prairie Press/Mischka Press, from Charles Philip Fox, *Working Horses: Looking Back 100 Years to America's Horse-Drawn Days: With 300 Historic Photographs.* 1st ed. (Whitewater, WI: Heart Prairie Press, 1990), 150. Photo credited to the Anna Fox Connection. (B) *Life,* June 1907, 757.
Note. (A) The caption accompanying the photo in Fox's book reads "A light spider phaeton with basket seat and rumble with English canopy presents very smart and fashionable equipage."

FIGURE 23. Broughams.

(continued on next page)

FIGURE **23.** (*continued*)

Sources. (A) *Life*, March 3, 1904, 200. (B) *Life*, January 19, 1905, 58 (inside front cover).

FIGURE **24.** Naming of vehicles.

Source. Photo courtesy Vermont Historical Society.
Note. This undated photo appeared in Charles Philip Fox and Jean Van Dyke, *Horses in Harness* (Greendale, WI: Reiman Associates, 1987), 118, with the caption "NAME THAT VEHICLE. Different manufacturers had different names for seemingly the same general vehicle. The above buggy might be referred to as a 'Road Wagon' or a 'Spindle Seat Driving Wagon', or, perhaps, a 'Runabout'. This vehicle the young ladies are enjoying near Waitsville, Vermont, has wire spoked rubber tire bike wheels."

technical discussions and the lengthy exchange of mechanical information between manufacturers and buyers;[156] it is possible that this practice evolved from the tradition of ordering buggy parts in which buyers needed to be fully acquainted with the inner workings of their machine.

At the same time, manufacturers purposefully promoted assimilation of the automobile by instructing the buying public to transfer preexisting practices from the horse culture to a new mechanical medium. A 1904 Pierce automobile advertisement talked of "the education of the automobilist" as a matter of transferring knowledge about horse buggies to automobiles, including criteria for purchase, such as price versus quality. With the predominance of cheap automobiles in the market during the first decade of the century, a Pierce advertisement attempted to render price a nonissue by emphasizing quality. The cost rationale remained

FIGURE 25. Runabout varieties.

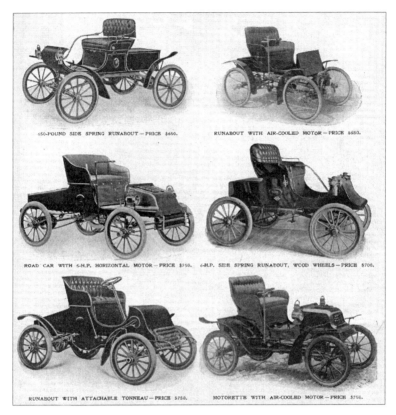

Source. The Automobile, 1903, 423.

equally compelling. Automotive advertisements instructed customers to buy a car as one would buy a horse and a buggy:

> A man buys an auto just as he formerly bought a buggy. The first time a man buys a buggy he buys a cheap one, because price is the only thing he understands about buggies. After he has had some experience, quality comes before price, and nothing but the highest grade carriage will satisfy him after that.[157]

At this time, a price war between horses and motorcars motivated many manufacturers. Many motor-buggy companies, particularly during 1907 and 1908, advertised their products as "cheaper than horses." The Lindsley delivery car, at $375, was claimed to have a 30-mile-per-gallon fuel efficiency, which meant that its operation was "much cheaper than horse help."[158] The Kiblinger, advertised frequently in 1907 and 1908 in *Collier's*, was claimed in large, bold print to be "cheaper than horses."[159] The Brush Runabout was advertised to "cost less to run than a horse."[160] A brand-new runabout in 1911 was claimed to be the same price ($350) as a good horse outfit.[161] Reliable Dayton, which also ran the same advertisement for many years in *Collier's*, claimed that its vehicle was "more economical and serviceable than a horse," and while it might not be lowest in price, it would be the "cheapest in the end."[162]

Rapid claimed that its commercial cars saved their owners anywhere from $500 to $1,200 per year.[163] A large Brooklyn, New York, department store was touted to have saved $1,360 during a six-month period by replacing its horse wagons with motor trucks for delivery.[164] The Metz Company published a customer testimonial in 1909 professing that their new motor vehicle "ha[d] taken the place of two driving horses" and that its upkeep cost was considerably less than that of maintaining one horse.[165] In the same year, Maxwell Motor Car advertised that its business runabout "costs less to keep than a horse and buggy" because "when not in use expenses stop," whereas "a horse eats all the time."[166] Some advertisements phrased it as a matter of the motorcar "not eating its head off when standing still."[167]

In 1910 Maxwell Motors printed a full-page advertisement entitled "Automobile vs. Horse" about a purportedly "disinterested" cost comparison between the automobile and the horse and buggy conducted under the supervision of the Contest Board of the American Automobile Association.[168] After several days of test runs for six hours a day, the board concluded that the automobile was cheaper to operate. The advertisement proclaimed,

> This is our answer to the charges made that the automobile is an extravagance. It proves the automobile an economic factor that would save millions if the Maxwell were everywhere substituted for the horse and buggy.[169]

Advertisers took the approach of showing how the automobile as a machine had to be taken *seriously* for its potential to "save millions" of dollars.

As late as 1916 the cost-justification campaign continued. The Federal Motor Truck Company, for instance, advertised that it reduced "horse-drawn haulage costs to fractions" when it replaced the "horse-drawn methods of yesterday" with motor power.[170] While these advertisements were geared to entice horse-and-buggy users, they implied that the automobile was the "new horse of today." By virtue of its association with the horse, the automobile came to justify its utility. By virtue of its cost-saving claims, it justified itself as a sensible replacement.

ESTABLISHING INTERCHANGEABILITY IN OPERATION

After considerations of functionality and cost, prospective buyers would now need to be convinced that the newfangled machines could be operated just like horse-drawn carriages. The Pope Manufacturing Company, famous for its bicycles, launched a Columbia motor carriage advertisement in 1898—one of the earliest[171]—claiming that "any one can learn [...] to control it in ten minutes."[172] In 1901 Toledo advertised its steam carriage to be so simple that "a woman can operate it conveniently."[173] That same year, Columbia advertised its cars to be "easily operated by man, woman or *child*" (emphasis mine).[174] As late as 1909 the Maxwell Motor Company advertised its gasoline cars as being so easy to drive that one's "wife, daughter or son" could run and care for it.[175] As late as 1913 the Haynes gasoline motorcar was proclaimed a children's product that "even women and children can drive [...] with enjoyment and safety."[176]

Thus, in addition to being economically sensible, the automobile was presented as being as easy to operate as a horse. Women and children had

FIGURE 26. *Life* illustration.

Source. *Life*, January 1, 1903, 9.
Note. Caption reads "Reggy's Christmas present."

been known to operate horseless carriages and thus, the argument was, if motorcars could be considered a replacement for horses, women and children could operate them as well. A 1903 *Life* illustration suggests that children, particularly the rich, received automobiles as presents and were allowed to drive them.[177] Illustrations of this type were published for several years, with pictures of gleeful children driving automobiles in the most dangerous circumstances (see Figure 26).[178]

A sixteen-year-old boy working for an automobile and repair company was given an official license to operate all classes of automobiles, including trucks, in 1903.[179] Three years later a fourteen-year-old boy was licensed as a chauffeur and made long trips around Fresno, California, during the busy fruit season.[180] Efforts were made to normalize automobiles and render them as harmless as horses such that women and young adults were deemed capable of controlling these machines.

An Oldsmobile advertisement of 1903 suggested that the controls for a gasoline vehicle required the same intuitive skills used to direct horses: "the controlling mechanism is simple, strong and instantly *responsive to the will of the driver*, giving a sense of perfect security" (emphasis mine).[181] While the horse is not explicitly mentioned in this ad, the automobile is given life in these descriptions—the unthinking motor suddenly becomes responsive to nonmechanical influences such as the "will" of the driver—it is no longer a "nerveless" machine. A similar description of a sentient machine appeared in a 1904 Northern automobile advertisement in which the motor was described to run smoothly and to give "instant obedience."[182]

While such descriptions would seem strange in modern-day advertisements, the early-twentieth-century sensibility was quite attuned to working with workhorses, thus the terms "obedience" and "will," befitting the training and handling of animals, were employed to make the motor seem more familiar. Indeed, manufacturers strained to make motor and muscle power operationally equivalent. Advertisements promoted automobiles as operating just like horse-drawn carriages.[183] The Autocar Company splashed its ad in *Life*, claiming its controls were "Simple As a Pair of Reins."

> The control of Type XI Autocar brings automobile driving to the simplicity of horse driving. In the rim of the steering wheel, and forming parts of it, are set two grips, one at the right hand, one at the left. These two grips control the throttle and the spark, regulating the speed of the car from 3 to 35 miles an hour. This arrangement brings the steering and the speed regulating together, so that in all ordinary running the only position necessary for the hands is on the steering wheel.[184]

The positioning of the accelerator on the steering wheel mimicked, although unsuccessfully, the experience of operating the reins of a horse-drawn carriage (Figure 27).

However, unlike the horse-drawn carriage, in which the effects of starter, brakes, accelerator, and steering wheel resulted from the manipulation of the reins, the motorcar had to separate these functions into different control mechanisms in order to mechanize actions horses perform. Indeed, as Harry Collins argues, intelligent work and movement had to be immensely simplified in order for it to be mechanized.[185] Levers and switches had to compensate for the sensibilities of the horse, particularly in braking and maneuvering. The driver had to pay close attention to the brakes, which constituted another lever. Nevertheless, advertisers strained to make the resemblance:

> To be sure this car has a gear shift lever and an emergency brake lever at the right of the driver. The gear shift, however, is needed only on particularly hard hills, or very bad bits of road. The car loaded with four passengers will climb nearly all hills on the high gear. The foot brakes being ample for all ordinary use, the emergency brake is rarely required. Hence we say that except in extreme cases only one position is required of the hands in driving Type XI Autocar.[186]

The concerted effort to downplay these compensating devices resulted from a desire to convince users that skills and mechanics employed in the driving of horses could be replicated in cars. However, ease in mechanical operation was not the same as riding a horse, although some pro-car advocates completely ignored these unmistakable differences.

FIGURE **27.** Steering wheels touted as reins.

Source. Life, September 7, 1905, 266 (inside front cover).

According to C. E. Woods, "The simplicity attached to the operation of an electric vehicle by any person of ordinary intelligence is too well-known to need comment at this point; but it is found from experience that there is the same difference in the care taken of an electric vehicle that there is among men who attend dynamos and steam engines, or drive horses, with a corresponding difference in troubles and aggravations."[187] Cars and horses, according to this pro-car advocate, demand the same level of conscientious maintenance. Those who do not adequately care for their means of transport should expect the same amount of problems in performance regardless of the type of power used.

A training course for automobile drivers[188] in 1899 strained to prove that operating a motorcar was like handling a horse. The driving school described the steering lever as similar to the steering of horse-propelled vehicles; it turned the rear wheels rather than the front. However, a more detailed description of automobile driving technique betrayed the school's effort to shift skills from horse- to car-handling. The school estimated a week of training (rather than 10 minutes) was required to "manage all the brakes and levers with perfect presence of mind."[189] The actual driving was described as follows:

> Both of his hands and both of his feet are fully employed. With his left hand he manages the power lever, pushing it forward one notch at a time to increase the speed. With his right hand he controls the steering-lever, which, by the way, turns the rear wheels and not the front ones, as is done with horse-propelled vehicles. His left heel is on the emergency switch, and his left toes ring the gong. With his right heel he turns the reversing-switch, and he can apply the brake with either his right or his left foot. When he wishes to turn on the lights, he presses a button under the edge of the seat. Hence, he is very fully employed, both mentally and physically. He can't go to sleep and let the old horse carry him home.[190]

This highly involved physical and mental activity—that is, the use of left and right heels to turn various switches, the use of the left toe to ring a gong, and so forth—constituted textbook driving. In practice, driving appeared to be even more complex and counterintuitive for anyone accustomed to riding a horse. A comprehensive article advising the novice on how to purchase a car, covering topics from price considerations to the various motor and style options, admitted that in the end it all came down to a matter of driving competence: "the great, essential fact that the beginner is apt to forget is that, for him, the first consideration is not to get a car whose power, size, and style meet every whim that he can devise, but *to get a car that he can run.*"[191] Driving ability, therefore, was a primary consideration. However, the skills required to drive a horse-drawn carriage turned out not to be easily transferable, if at all.

A seasoned car driver described the many difficulties of learning the various nuances of early cars, such as the lack of an efficient means to prevent a car from running backward if stopped on a hill, even with a first-class machine such as the Daimler.[192] He learned from experience that when a car ran backward, the best way to make it move forward was actually to put the reverse gear into action.[193] Clearly such peculiarities found in machines operating on actual roads and in traffic provided another learning gap that a driver had to quickly fill in as he drove along. The qualifications made by the earlier Type XI Autocar advertisement regarding the rare use of the gearshift lever and emergency brakes were clearly understated, particularly since "bad" roads were the norm at that time.

The greater demand on the driver's mental and physical faculties was such that some companies built engines that could be operated with reins in order to facilitate the transition from driving horses to motorized tractors.[194] The advertisement for the Klaxon horn, for instance, described the "pressure of [its] button" as warning horses to keep them at bay "as surely as if you held in your own hands the reins of every horse ahead."[195] This effort to connect mechanical switches, buttons, and levers directly to horses and reins signifies a purposeful effort to apply equine concepts and practices to the motorcar. A slew of advertisements strained to make the newfangled machine interchangeable with the horse in form and functionality.

Owning an Entire Stable of Horses in One Vehicle

Automobile advertisers took every opportunity to describe motor performance in terms of the work traditionally performed by horses. As early as 1901 the Triumph Motor Vehicle Company advertised that its automobile "climb(s) any hill a horse can climb" and that it "carries two people, and will go over any road a horse could travel on."[196] This direct association with horses could also be seen in an advertisement for an Oldsmobile whose range of applications was posted simply as "an entire stable of horses in itself."[197] The Reliable Dayton Motor Car, for which the same ad ran for at least three years from 1907 to 1909, was claimed to be "the First Real Successor to the Horse" because it had "the same reliability as the horse."[198] The Herreshoff car in 1909 was described as "the smart, light cob of the automobile stable."[199]

These direct appeals to the functionality and capabilities of the horse provided advertisers with the means to orient consumer perception. Advertisers presented their automobiles as able to accomplish any function performed by horses, with faster and better results. A more explicit comparison between literal horse power and figurative horsepower came from a 1905 advertisement of the Autocar Company:

> The Auto Runabout has a motor of 10 mechanical horsepower which is about equivalent to 15 animal horse-power.
> The horses, however, could furnish this power for only 8 hours a day. As the Autocar Runabout can be run for 24 hours a day, it will be seen that if both power and endurance are considered this car is equal to 45 horses.[200]

In translating mechanical horsepower in terms of animal horse power, advertisers sought to display in quantifiable terms the superiority of motor performance over that of horses. Kiblinger Motor Wagons, which often advertised in the front pages of *Collier's*, sometimes right under the editorial bulletin, claimed that not only would its machine "climb hills easier than horses" but also that it would "do quicker work than 3 horses and wagons."[201] A 1908 advertisement for the Lindsley claimed that it would "easily do the work of 5 delivery wagons of the horse kind."[202] Rapid recommended "discarding the antique method of horse-drawn vehicles" and instead "install[ing] motor wagons which cover 3 to 5 times as much territory as horse-drawn wagons, and give less trouble as they never tire, are easy to operate, and cost less to maintain."[203]

The message was clear, consistent, and definitive: motorcars worked just like horse-drawn carriages but were far superior in performance and economy. Automobiles became continuously associated with horses while at the same time jostling to replace them. On one level, horses were expensive, inefficient, and weak, but on another, motorcars worked just like horses. Manufacturers used the horse to explain the automobile concept but at the same time discounted the horse in order to justify replacing it.

This tension between continuity and discontinuity relates to the strength of preexisting practices in defining the admissibility of new devices. Manufacturers recognized the entrenched position of the horse in people's work routines and work processes. People carried out their work in terms of their conceptualization of the horse. Thus the automobile as a nascent technology could not redefine work and social routines by presenting itself as being unlike the horse; instead it needed to prove itself a much "better horse."

RECOGNIZING WOMEN'S PATRONAGE

A marked shift occurred, an observer remarked in 1905, when the automobile shed its "machine-like" qualities in order to become more "coach-like."[204] As late as 1923 advertisements such as that of Delco Electrical Systems still called automobiles "A Stage-Coach of Today" and considered them "modern successors of the old stage coaches."[205] The move toward a coachlike effect in automobile construction came as early as 1900, when Woods Motor Vehicle advocated the car for personal transport rather than for sport, describing its motor carriage as the "lightest, smartest-looking and most graceful Automobile ever built."[206] Several years later advertisers continued to emphasize the coachlike image of their offerings—their "luxurious upholstering and elegant finish," "refinement in design," "perfect taste," and "essence of good form"—descriptions typically found in many electric vehicle advertisements, such as a 1908 ad for the Thomas Town Car.[207]

Gasoline vehicles, on the other hand, emphasized the riding qualities of coaches. A 1909 Oldsmobile advertisement promoted large wheels as a protection against ruts, cracks, or any "inequality of the road" that might compromise luxurious riding.[208] The Brush Runabout in 1908 was claimed to "be durable, certain, comfortable, lively, handsome, almost noiseless, almost vibrationless and [to ride] like a baby carriage."[209] The more definitive "no vibration, noise or odor" description of the Riker Motor Vehicle, sometimes phrased as "if you appreciate in an Automobile Cleanliness, Freedom from Noise, Vibration, and Odor, you will buy a Riker,"[210] hinted at some of the public-relations battles manufacturers faced in repackaging the automobile.

The transformation of the automobile from its early days of reckless joyriding—when the various sights, sounds, and smells it made repulsed the American public—required a shift in purpose from the driving to the riding experience. Focusing on passenger comfort included addressing nonvisible concerns, such as fears and apprehensions about the unruly machine. Advertisers such as Lindsley typified this effort, claiming that its car was "safe, *sane*, practical and durable" (emphasis mine).[211] The terms "safe" and "sane" appearing in automobile advertisements suggest an anthropomorphic but mad machine that manufacturers must domesticate. The actual riding experience, as exemplified by the Lindsley advertisement, was assured to be suitable "to those ordinarily nervous when in a motor car," and the car to be "the safest car for women's use."[212]

The focus on women as a customer base constituted an effort to associate cars with comfort and safety.[213] The functional use of the car in the 1904 Pope-Waverley advertisement listed female-related purposes first and foremost, claiming that "ladies prefer them [automobiles] for shopping and calling," while business reasons come second, with appeals to physicians based on their "readiness and economy" (Figure 28).[214] The dependence of the nascent automobile industry on women's patronage would later become apparent.

FIGURE **28.** Pope-Waverley Electric Car advertisement.

Source. Life, June 2, 1904, 523.

One could argue that automobiles still served entertainment purposes at this time, but the emphasis was on the ability of the car to take its driver *somewhere* to be entertained, rather than on the driving experience itself as entertainment. A 1904 Oldsmobile advertisement promised to take a woman to an enjoyable destination: the Oldsmobile was an "ideal machine for any woman who enjoys the outdoor life" because it was "safe, reliable, easy to operate" and would bring her in perfect comfort to her destination, "rain or sunshine."[215] The promise of a delightful afternoon was found in places, people, and various activities rather than in the automobile itself, signifying a move away from its sporting past and perhaps a move toward mainstream status.

The taming of the devil wagon can be understood in terms of simultaneous equine and feminine articulations. The move toward projecting a coachlike appearance required the motorcar to be the equivalent of the horse carriage not only in its functionality but also in its virtuous qualities. Franklin Motor Cars described its steam vehicles in 1904 as being "as sensitive and spirited as a thoroughbred horse."[216] The emphasis on cars having the genteel qualities of thoroughbred horses appeared in many other earlier buggy advertisements, such as that of the 1899 Spider Stanhope. Stanhope advertised the "perfectly correct" and harmonious style of its carriages and again used only women as drivers and passengers in its posters.[217]

A Franklin Motor Car advertisement also featured a woman at the controls of its steam vehicle, with a man seated on the passenger side. Many automobile advertisements featured female drivers and passengers exclusively. The background undertones were feminine and evoked the imagery of a carriage—"light, flexible, ease of management, and extreme luxury" to replace the "unnecessary, heavy and troublesome" mechanical contraption of the past.[218] Although a newer machine was in fact replacing an older one, the imagery played on familiar domestic themes rather than on technical improvements. The newer machine was in fact ushering in old, familiar sentiments.

The use of women as a domesticating device enhanced the automobile's coachlike reputation because women conveyed the notion of safety: it was observed that women were careful drivers and thus generally not involved in vehicle accidents.[219] Feminine descriptors intertwined with equine features appeared consistently in many advertisements in the early twentieth century regardless of the type of motor power. During this period Haynes-Apperson used women almost exclusively in many of its advertisements for its gasoline automobiles in *Life* and *Collier's*. Woods Motor Vehicle similarly used only women in its advertisements.[220] The extensive use of women in advertisements in the early twentieth century constituted a different sensibility from their use in modern-day automobile advertisements.

Women in the early twentieth century were considered an influential consumer base and thus were depicted as potential customers, whereas modern-day advertisements tend to use women decoratively.[221] In the advertisements of the early twentieth century, women were depicted in the driver's seat and were sometimes illustrated from a distance. Modern-day advertisements, on the other hand, show women as what Scharff calls "automotive accessories," exemplified by a 1948 photo in her book in which a woman is used as a hood ornament.[222] Other, more popular advertisements show women flanking automobiles in close-up, provocative full-body poses, but rarely operating them.

The early woman automobile shopper, on the other hand, was described as "out in force" by 1906: she "look[ed] over the automobile advertisements in the daily newspapers and ma[de] the rounds of the salesrooms in about the same frame of mind that she would if she were shopping for a new bonnet."[223] Some observers believed that the entire automobile industry in its early years relied on women's patronage for its survival. As early as 1898 a writer noted the manner in which women had taken an interest in the new machine:

> The women, strange to say, have from the first shown more enthusiasm than the men for the new vehicle. It is they who have been its most ardent promoters, who have organized horseless coaching excursions, given prizes for races—in short, created the first paying demand for a clean, speedy and reliable machine.[224]

The notion that women created the "first paying demand" for a practical automobile problematizes the commonly held gendered typifications of the car as a male domain that came to be usurped by women. Scharff, for instance, argues that "the auto was born in a masculine manger, and when women became drivers, they had to overcome their own lack of confidence and combat both subtle and overt resistance."[225] Women supposedly appeared in many advertisements as passive figures seated next to a man.[226]

Historical data, however, show less timid and, in fact, highly participatory behavior among women. In 1901 an automobile writer wrote that women—"plenty of them"—were already driving their own automobiles.[227] A 1901 advertisement in *Collier's* described the Triumph automobile as a "swell carriage" for ladies without mentioning other prospective customers.[228] In 1903 Oldsmobile, Haynes-Apperson, Searchmont, Cudell, and many other manufacturers used women almost exclusively in many of their advertisements. Even tire companies such as the Hartford Rubber Works Company generally depicted a woman behind the wheel during the years 1903–1904 (Figure 29).[229] As late as 1908 Stepney Spare Wheel tires also used women in their advertisement in *Collier's*.[230]

In 1906 the *Ladies Home Journal* featured the story of an enterprising young woman who organized a two-week excursion for 24 girls, including a chaperone, for a trip from Philadelphia to the Delaware Water Gap for only $1.60 per day including expenses.[231] By this time, many women easily handled large touring cars, and a significant number of them owned and operated their own automobiles.[232] It was not uncommon in 1906 to see large touring cars filled with women driving in congested business districts and on city boulevards.[233] Four women made a transcontinental trip in a Maxwell car in 1909, performing all the necessary repairs themselves along the way.[234] By this time, a column written specifically for women drivers dispensed advice on many aspects of driving including how to handle emergencies and avoid accidents.[235]

While Scharff acknowledges that a few gasoline auto manufacturers recognized a female market for their products,[236] she generally implies that most manufacturers did not specifically target women drivers for

Figure 29. Dunlop Tire advertisement.

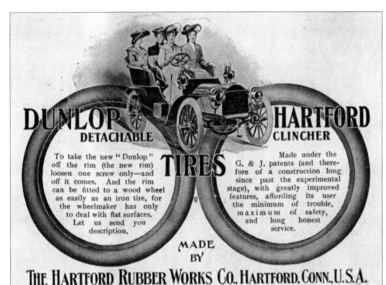

Source. *Life*, April 7, 1904, 351.

their larger gasoline touring cars. Scharff claims that the rapidly grow-ing number of women driving gasoline cars disrupted the gendered stereotype in the minds of the advertisers who saw electric cars as *the* car for women and gasoline cars as the car for men.[237]

However, advertisements depicted women at the helms of gasoline cars as early as 1902 (Figure 30). The emphasis on the ease of operating gasoline cars at this early stage confirms the idea that manufacturers would sell a car to anyone, regardless of gender. Indeed, whether manufacturers were catering to the needs of women or specifically targeting them as potential buyers could be argued in a number of differ-ent ways, but one thing is certain: manufacturers had women foremost in their minds from the start (Figures 30 through 39).

John W. Haynes, a well-known race-car driver, noted in 1907 that women "who learn to drive cars are as a rule exceptionally capable after they have mastered the mechanical details of the work."[238] Many women automobile workers also helped commercialize cars. It was observed in 1909 that women performed the "better class of work" in almost every automobile plant, particularly for tops and upholstery.[239] The notion of the car as being "born out of a masculine manger," as Scharff argues,[240] may require further qualification, particularly when the birth of the automobile appears to be intimately tied to the patronage of women as buyers and as factory workers.

FIGURE **30.** Haynes-Apperson advertisement.

Source. Life, March 6, 1902, 182.

FIGURE **31.** Peerless advertisement.

Source. Life, August 6, 1903, 119.

FIGURE 32. Searchmont advertisement.

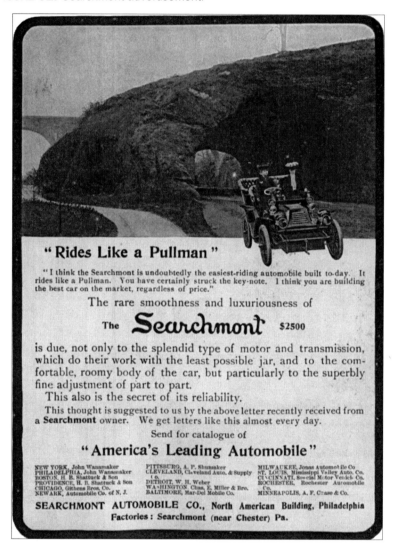

Source. *Life*, July 23, 1903, 93.

FIGURE **33.** Cudell advertisement.

FIGURE 34. Franklin advertisement.

FIGURE **35.** Studebaker advertisement.

Source. Life, June 2, 1904, 560.

FIGURE 36. Studebaker advertisement.

Source. Life, May 5, 1904, 446.

FIGURE **37.** Locomobile advertisement.

Source. McClure's, April 1905, 75.

FIGURE 38. Oldsmobile advertisement.

Source. *Life*, April 6, 1905, 360.

FIGURE **39.** Pope-Waverley advertisement.

Source. Ladies Home Journal, July 1906, 27.

Although there is little accurate data on the exact number of drivers in the first decade of the twentieth century,[241] female motorists certainly were featured in many accounts of motoring. A noted automobile observer in Buffalo, New York, reported that women drove to the shopping district in their own cars, paying no heed to the cold or snow, whereas three or four years earlier no one, including men, had dared to drive in Buffalo's winter weather.[242] As early as 1903 women in many cities, such as Pittsburgh, demanded a separate motor school and instructors of their own.[243] At this time, the Ladies' Automobile Club of Great Britain and Ireland secured a large drawing room in the Hans Crescent Hotel for six months reserved exclusively for lectures and practical lessons in motorcar driving.[244] The Automobile Club of America's annual banquet menu card in 1903 featured a woman on its cover.[245] In 1902 *Life* began publishing advertisements for automobiles featuring women as passengers and drivers.

It is conceivable that manufacturers of electric cars focused specifically on *wealthy* women rather than on women in general. Thus the contrast between electric and gasoline as gender-driven suggested by Scharff could be argued along the lines of economic and class distinctions. Yet even the class issue does not seem to provide a neat classificatory scheme for the various types of motors. A 1903 Oldsmobile advertisement in *Life*, for instance, attempted to appeal to economic prestige by showing the continuity of upper-class mannerisms despite the change in transport (Figure 40).[246] An almost identical image appeared in *Life* seven years later that depicted two wealthy women, one from 1710 and the other from 1910, meeting on common ground despite having arrived in different types of transport.[247]

Perhaps rather than viewing different types of automobiles as mechanisms for creating different types of gender and class distinctions, one could examine the availability of different price and model options as indicators of the kinds of customers the automobile attracted. Class distinctions in existence during the horse-and-carriage days appeared to have been transplanted into the motor era: specific cars were designed for specific types of buyers.

FIGURE 40. Oldsmobile advertisement.

Source. Life, September 3, 1903, 210 (inside front cover).

The Automobile's comprehensive profile of car models available in the market in 1906 showed that both the *lowest* price category, from $400 to $800, and the highest price category, from $6,500 to $12,300, consisted of gasoline cars.[248] Gasoline cars dominated both ends of the price spectrum, from most affordable to most luxurious. Manufacturers were keen on catering to a broad customer base. Electric cars also competed with gasoline cars in virtually all price categories.

During the horse era, some carriages were built specifically for the upper class, but the automobile era also appealed to the same class distinctions regardless of the motor power used. A 1915 *Life* illustration best exemplifies how the social practice was simply transferred to an automotive medium (Figure 41).[249]

FIGURE 41. *Life* illustration.

Source. *Life*, May 13, 1915, 861.
Note. Caption reads "Yes, she's a typical nouveau riche. Can't trace her ancestry farther back than 1890."

Nearly all automobile builders at that time admitted to studying idio-syncrasies in fashion, particularly with regard to color, in order to effec-tively harmonize or contrast cars with women's dresses.[250] A company that made one of America's costliest cars sold two vehicles at a private exhibition to two women buyers who bought the cars simply on the basis of their being the right color to match their wardrobes.[251] Many socialites had their cars repainted regularly to match their gowns.[252]

SUPPLANTING THE HORSE

The transplantation of social practices onto the automotive platform occurred as manufacturers sought to court mainstream users. While early advertisers may have adopted persuasive measures to convert early horse-and-buggy users, early adopters themselves informed advertisers in various creative ways to make this transition work. For instance, as early as 1896, for a horseless carriage race conducted on the horse track in Narragansett Park at the Rhode Island State Fair, all drivers wore jockey outfits of derby shirts and hats rather than goggles and hoods.[253] The sport of foxhunting in Nashville, Tennessee, had been greatly facilitated by the use of the automobile in 1905 because it decreased travel time to and from the hunting grounds.[254] Seven years later, the hunt itself in

England was described as having been "revolutionized" by the replacement of horses with automobiles.[255]

A telling *Life* cartoon captioned "Will the horse be superseded by the automobile in every equestrian sport?" featured a woman behind the wheel controlling a series of cars harnessed together like horses (Figure 42). Almost thirty years later, a similar analogy appeared in an advertisement showing a man standing beside a car and a woman sitting behind the wheel: "Got any tips, Tommy?" asks the woman. "Well, they say Kelly is a good bet in the third race," replies the man. The woman answers, "That sounds good to me. I'd bet on a Kelly any time."[256]

Thus there were cases when suggestions of a direct replacement of the horse were made without any modifications of practices. Perhaps a more telling example of the enduring character of preexisting practices is the picture of a woman in *The Automobile Magazine* in 1902, her spinning wheel replaced by a steering wheel, with the caption "the more things change, the more they stay the same" (Figure 67).[257]

Tales of courtship with lovers eloping via automobile instead of horse were a common theme. For instance, an illustrated story of a father running after his daughter showed young lovers escaping in an automobile while the father tried to catch up with his horse-drawn

FIGURE 42. *Life* illustration.

Source. Life, January 7, 1909, 23.

carriage.[258] Some stories used "then" versus "now" pictures showing lovers running away in a horse-drawn carriage with Cupid in the background pointing his arrow at the lovers ("then") only to be replaced with an automobile ("now") with Cupid chasing after them on a bicycle (Figure 43).[259]

Such illustrations suggest that nothing much had changed except for the direct replacement of the horse with the automobile and the obvious difference in speed. These reassuring and familiar images surrounding a new and increasingly proliferating technological device could perhaps be taken as a cynical social commentary on the inability to change with the times. Nevertheless, the emphasis on continuity of practices associated with the horse in advertisements and social commentaries assures the consistency of the message given to the public—using an automobile would not radically change people's lives.

FIGURE **43.** *Life* illustration.

Source. Life, February 4, 1909, 163.

Suggestions of automobiles supplanting horses were also depicted in traditional folklore. For example, an illustration of Lady Godiva riding in an automobile rather than on a horse ("Had Lady Godiva Lived To-day"; Figure 44)[260] attests to the many ways horses were physically supplanted by automobiles, although not in conceptual and functional terms. In more practical matters, street sprinklers, ambulances, and fire trucks were motorized in the first decade of the twentieth century. The general appearance of these public service vehicles has been described as closely resembling their horse-drawn equivalents, with the motor doing the work of the horse.[261] American Indians, or "aboriginal Americans" as they were called in the early days, were said to have caught the motor craze as well, with an *Automobile* cover article claiming that the "noble red man seems to have [taken] very kindly to the automobile" and that a number of them owned and operated their own machines on the reservations.[262] A 1909 cartoon depicted a cowboy driving an

FIGURE 44. *Life* illustration.

Source. Life, September 13, 1906, 281.

automobile in pursuit of a cow much as he would if he were riding a horse (Figure 45).[263]

California ranchers were described as using motorcars to ride up to "snow peaks and down to desert valleys below the level of the sea," as one observer claimed; automobiles were supposedly as common as horses in sparsely populated areas of the cattle country of the West from as early as the first decade of the twentieth century.[264] These later examples again show the increasing replacement of the physical horse but the retention of its functional and conceptual uses.

The automobile also made use of preexisting distribution and service infrastructures built around the horse. Commercial supply routes in urban areas relied on horses for the distribution of goods and services. According to McShane and Tarr, the entire internal city circulatory system at the turn of the twentieth century, such as freight delivery, passenger transportation, food distribution, and police and ambulance services, was constructed around the horse.[265]

Many carriage companies such as Pritchard, Edwards and Sullivan[266] in 1890 began selling horseless carriages side by side with buggies in

FIGURE **45.** *Life* illustration.

Source. Life, March 4, 1909, 287.
Note. Caption reads "The latest style in cow-boys."

their stores; this continued into the early twentieth century. The Fiss, Doerr & Carroll Horse Company of New York City took on the exclusive distributorship of Columbia electric vehicles.[267] The Studebaker Carriage Company advertised automobiles and horse carriages side by side in many of its posters.[268] Sales barns for horses were used to exhibit automobiles.[269] Secondhand cars were sold on horse auction blocks.[270] Storage facilities were referred to as "automobile stables" rather than garages.[271] Terms such as a "stable full of cars," used in many Oldsmobile and Herreshoff advertisements, continued to appear as late as 1909. Shifting the infrastructure of carriages to cars occurred with ease and without massive start-up costs.

While horses had been argued to be a "liability" compared with automobiles, particularly in terms of daily operation and maintenance for public transport,[272] the rapid diffusion of the automobile appears to have been facilitated by the ease with which it inserted itself into well-established infrastructures of the horse culture such as distribution and sales systems and service and maintenance centers. In positioning itself to work like a horse, the automobile did not have to challenge the existing work structure directly in order to find its way into people's lives. It merely had to supplant the horse.

CONCLUSION

The momentum gained in shifting from muscle to motor power required mimetic measures. Motorized power posing as muscle power came as a move to pull the automobile from the periphery to the very center of society—automobile manufacturers employed various means to project an image of a reliable, easy-to-operate, mundane, practical everyday form of transport in order to overcome an infamous reputation of technical novelty with dubious functionality. Early automotive advertisements provide interesting insights into how this was accomplished.

During the first decade of the century, automotive manufacturers had to contend with uncertainties about the viability of a self-propelling vehicle and its ability to deliver benefits not already being provided by the horse. The automobile had to shed its whimsical and sporting image to be accepted as a form of everyday transport. Despite its significant differences from the horse in terms of driving, operational controls, and, for that matter, sights, smells, and sounds, the automobile was presented as the horse's operational (motorcar steering working like reins) and functional (working just like a horse) equivalent. As a writer observed in 1905, there was "in automobile affairs the constantly recurring allusion to the work of horse-drawn vehicles."[273]

In becoming like a horse, the automobile justified itself as a useful device. At the same time, in becoming *unlike* the horse—a cheaper and better alternative—it justified itself as a replacement. An interesting paradox transpired in many motorcar advertisements. Horses were supposed to be inferior, needing to be replaced with machines—machines that worked just like horses. Hence, the intentional effort to align the automobile with horse culture raises the question of technical superiority and economic sensibility as the sole determinants of automotive diffusion. I argue that by virtue of its association with the horse, the automobile was transformed from a deviant spectacle into a commonplace device of everyday life. The automobile, loathed by the public, came to assume the mantle of familiarity and conformity.

The transformation of the automobile into a familiar mode of transport also came to be understood in terms of feminine articulations. Women constituted the "first paying" customers who insisted on a machine that was clean, speedy, and reliable.[274] The use of women exclusively in many early automotive advertisements showed their importance as patrons and had the added benefit of conveying an image of safety, comfort, and ease of use. Feminine descriptors intertwined with equine features helped to enhance the automobile's reputation as a replacement for the horse-drawn carriage.

Women contributed not just in the consumption but also in the production of early automobiles. Women were described to perform the "better class of work" in almost every automobile plant, particularly for tops and upholstery.[275] Automotive manufacturers hired carriage mechanics and builders to create a more coachlike automobile, emphasizing comfortable ride over exhilarating drive, bringing in another familiar element of the horse-drawn culture to mainstream commuters. Traditional names and designs of horse-drawn carriages were copied.

The need for motor power to be interchangeable with muscle power may appear to have emerged seamlessly from the strength of preexisting practices rooted in the horse culture. However, it appears that this transition, facilitated by manufacturers and early adopters in response to public concerns, was made palpable through advertising and print media. Automobile advertisements in particular borrowed heavily from preexisting social norms concerning the various uses of the horse to redirect consumer orientation from literal horse power to figurative horsepower. The horselessness of the motorcar, its novel aspect, was less emphasized. Rather, the motorcar was simply introduced as a functional equivalent of the horse without its organic limitations. The motorcar co-opted preexisting equine culture and practices in order to survive, particularly in its early and most precarious stage of development.

Automotive ubiquity did not occur simply because cars became much more affordable and cleaner and performed more efficiently than horses. Historical studies generally have not addressed how the automobile relied on society's long-standing working relationship with the horse in order to be understood and accepted. When cars began to be seen as functioning like horses, rapid diffusion ensued as socioeconomic and cultural practices built around the horse were transferred to the car.

In the next chapter, I will show how this recurring dependency of the motorcar on equine culture persisted even in a military setting. In an attempt to motorize its cavalry, the U.S. military was caught in a similar predicament of having to preserve its long-held tradition of battlefield speed and mobility, which had been carried out using the horse. At the same time, there was urgent pressure to motorize, and a battle ensued between the proponents of the horse and the proponents of the motorcar.

While the motorcar appeared to have gained some military support from a policy standpoint after World War I, motorization was far from straightforward and, in fact, faltered despite direct mandates from those in high military offices. Reconciliation occurred when specifications for a reconnaissance car followed the performance requirements of the horse, paving the way for the most popular vehicle in World War II, the jeep. However, this reconciliation occurred at a slow and arduous pace.

Endnotes

1. For details on the car culture of the United States, see James J. Flink, *The Car Culture* (Cambridge, MA: MIT Press, 1975).
2. John B. Rae, "The Rationalization of Production," in *Technology in Western Civilization*, ed. Melvin Kranzberg and Carroll W. Pursell, Jr. (New York: Oxford University Press, 1967), 49.
3. Rae suggests that Europe's lack of market demand failed to sustain the English "rationalization" program or Germany's "Fordismus" (Ibid.). Presumably, other automotive infrastructure systems found in industrial America at this time, such as the distribution and repair service systems, did not materialize in Europe precisely because of the absence of market support. Thus, if a market does not have sufficient demand capacity, mass production will not materialize or, at least, be sustainable, as exemplified by Europe in the 1920s–1930s.
4. Frank Luther Mott, *A History of American Magazines 1885–1905* (Cambridge, MA: Harvard University Press, 1957).
5. Ruth Schwartz Cowan, "The Industrial Revolution in the Home," in *The Social Shaping of Technology*, Donald MacKenzie and Judy Wajcman (Buckingham, U.K.: Open University Press, 1999), 296.
6. Richard Ohmann, *Selling Culture: Magazines, Markets, and Class at the Turn of the Century* (London: Verso, 1996), 72–74, 100.
7. Kenneth MacKarness Goode, "Ten Years After: A Review of the Automobile Industry to the Present Day," *Collier's*, November 2, 1907, 14.
8. Ibid.
9. Pamela Walker Laird, "The Car Without a Single Weakness: Early Automobile Advertising," *Technology and Culture* 37, no. 4 (October 1996): 797. She adopted the idea from Donald Finlay Davis, *Conspicuous Production: Automobiles and Elites in Detroit, 1899–1933* (Philadelphia: Temple University Press, 1988), esp. 1–3, 20–25.
10. Laird, "The Car Without a Single Weakness," 797.
11. Ibid.
12. Ibid.
13. Ibid., 801.
14. Ibid., 797–98, again adopted from Davis, passim.
15. Ibid., 801.
16. Ibid.

17. Ibid.
18. Ibid., 797.
19. Virginia Scharff, *Taking the Wheel: Women and the Coming of the Motor Age* (New York: Free Press, 1991), 36–37, 44, 60, 87, 123.
20. Mott, *A History of American Magazines*, 28.
21. Matthew Schneirov, *The Dream of a New Social Order: Popular Magazines in America, 1893–1914* (New York: Columbia University Press, 1994), 78.
22. James J. Flink, *America Adopts the Automobile, 1895–1910* (Cambridge, MA: MIT Press, 1970), 1.
23. Charles E. Duryea, "As It Was in the Beginning," *The Automobile*, January 7, 1909, 47.
24. Ibid.
25. Ibid.
26. Elwood Haynes, "A Few Reminiscences of the Early Automobile," *The Horseless Age*, December 27, 1911, 957.
27. Schneirov, *The Dream of a New Social Order*, 184.
28. Ray Stannard Baker, "The Automobile in Common Use," *McClure's Magazine*, July 1899, 195.
29. Haynes, "A Few Reminiscences," 957.
30. Goode, "Ten Years After," 12.
31. Cleveland Moffett, "The Edge of the Future," *McClure's Magazine*, July 1896, 153.
32. Hermann F. Cuntz, "The Automobile as a Feeder of Civilization," *The Automobile*, June 10, 1909, 952.
33. Goode, "Ten Years After," 12.
34. R. H. Thurston, "The Automobile or Horseless Carriage," *Collier's*, April 28, 1900, 9.
35. "The Contributor's Club," *The Atlantic Monthly*, December 1901, 863.
36. Herbert L. Towle, "The Coming of the Automobile," *Collier's*, January 12, 1901, 33.
37. Reliable Dayton Motor Car ran the same basic advertisement in *Collier's* from 1907 to 1909.
38. "The Selection of a Gasoline Automobile According to Price," *The Automobile*, April 18, 1903, 422.
39. "Cars Offered For the Season of 1905," *The Automobile*, January 14, 1905, 49–63.
40. "Chief Characteristics of 1906 Models," *The Automobile*, January 11, 1906, 32–66.

41. *Encyclopedia of the U.S. Census*, ed. Margo J. Anderson (Washington, DC: CQ Press, 2000), 260.
42. Scott Derks, ed., *The Value of a Dollar: Prices and Incomes in the United States, 1860–1999* (Lakeville, CT: Grey House, 1999), 74.
43. Ibid.
44. "Cars Offered for the Season of 1905," 49–50.
45. *The Value of a Dollar*, 73.
46. *Collier's*, 1907–1908.
47. "The Poor Man's Automobile," *The Horseless Age*, July 12, 1911, 49.
48. Motor oil cost approximately sixty cents per gallon in 1908. See Scott Derks, *The Value of a Dollar: Prices and Incomes in the United States, 1860–2004* (Millerton, NY: Grey House Publishing, 2004), 93.
49. *Collier's*, 1908.
50. "Cars New and Second-Hand," *The Automobile*, August 3, 1905, 147.
51. Herbert L. Towle, "The Best Car for the Novice?," *The Automobile*, October 12, 1905, 395.
52. A. V. A. McHarg, "A Story of the Cry of 'Get A Horse,'" *The Automobile*, January 16, 1908, 79.
53. Flink, *America Adopts the Automobile*, 53. See also Flink, *The Car Culture*, 35, and Flink and American Council of Learned Societies, *The Automobile Age* (Cambridge, MA: MIT Press, 1988), 138.
54. Duryea, "As It Was in the Beginning," 47.
55. "The Horseless Carriage," About the World, *Scribner's Magazine*, March 1896, 393.
56. C. E. Woods, *The Electric Automobile: Its Construction, Care, and Operation* (New York: H. S. Stone & Company, 1900), 31.
57. "Survival of the Horseless Carriage," *The Automobile*, August 21, 1906, 154.
58. *Life*, September 25, 1913, 533.
59. B. D. Woodward, "The Exposition of 1900," *The North American Review*, April 1900, 476.
60. Goode, "Ten Years After," 12.
61. Henri Dumay, "The Locomotion of the Future," *Collier's*, July 30, 1898, 22.
62. M. M. Musselman, *Get a Horse! The Story of the Automobile in America* (Philadelphia: J. B. Lippincott Company, 1950), 166.
63. *Life*, June 19, 1905, 68.
64. Marquis De Chasseloup-Laubat, "Recent Progress of Automobilism in France," *The North American Review*, September 1899, 411.
65. Ibid., 406–407.

66. George Fitch, "The Automobile," *Collier's*, September 19, 1908, 27.
67. R. H. Thurston, "The Coming Automobile," *Collier's*, April 27, 1901, 9.
68. A tube used to generate an electric spark to ignite an air-fuel mixture that produced the motive force.
69. Dawson Turner, "Some Experiences With Modern Motor-Cars," *The Living Age Company*, June 9, 1900, 638–642.
70. "The Next Exhibition of Automobiles" [in French; adapted from *Journal Amusant*], *Life*, August 4, 1898, 99.
71. *Life*, August 6, 1903, 121.
72. Moffett, "The Edge of the Future," 156.
73. Air reduced in volume and held under pressure used to operate mechanical devices such as brakes.
74. Fitch, "The Automobile," 27.
75. Turner, "Some Experiences With Modern Motor-Cars," 641.
76. "Motor Carriages" [from *The Horseless Carriages* by Prof. John Trowbridge], *The Living Age Company*, April 10, 1897, 131–132.
77. Wolfgang Schivelbusch, *The Railway Journey: The Industrialization of Time and Space in the 19th Century* (Berkeley: University of California Press, 1986), xiii.
78. "How To Be Hideous," *The Automobile*, July 11, 1903, 29.
79. "An Authority," "Automobile Fashions for Women: Newest French and American," *The Automobile and Motor Review*, November 1, 1902, 12.
80. Ibid.
81. Walter Wellman, "Faster than the Express-Train: The Automobile Race from Paris to Berlin," *McClure's Magazine*, November 1901–April 1902, 21.
82. *The Automobile*, August 17, 1905, [no page number; adapted from *Sans-Gene*].
83. "Decay of Speed Laws and Ordinances," *The Automobile*, November 14, 1903, 499.
84. "Imperfect," *Life*, November 21, 1901, 415.
85. Ibid.
86. *Life*, July 17, 1902, 49.
87. *Life*, July 9, 1903, 34 (inside front cover). Also in *Life*, July 23, 1903, 94.
88. *Life*, July 30, 1903, 103.
89. "Automobilist Sues Policeman," *The Automobile*, July 4, 1903, 20.
90. *Life*, May 12, 1904, 459.
91. Walter Camp, "The Automobile," *Collier's*, March 9, 1901, 21.
92. Julian Street, "The Fools at the Finish," *Collier's*, November 7, 1908, 16.

93. Charles Belmont Davis, "The First Man Back," *Collier's*, November 7, 1908, 23.

94. Goode, "Ten Years After," 14.

95. "Decay of Speed Laws and Ordinances," 499.

96. "All Pittsburgh Aroused," *The Automobile*, August 10, 1905, 174.

97. "Reckless Chauffeurs Arrested," *The Automobile*, January 21, 1905, 158.

98. Ibid.

99. Kevin Borg, "The 'Chauffeur Problem' in the Early Auto Era: Structuration Theory and the Users of Technology," *Technology and Culture* 40, no. 4 (1999): 797.

100. Ibid., 802.

101. Ibid., 806.

102. Frank A. Munsey, "The Automobile in America," *The Automobile*, February 1, 1906, 313.

103. "Why Not Here?," *Life*, August 17, 1905, 205.

104. De Witt C. Morrell, "Reckless Employers" [in "Comments and Queries of Readers"], *The Horseless Age*, December 8, 1909, 652.

105. Borg, "The 'Chauffeur Problem,'" 814–815.

106. Ibid., 816.

107. Ibid., 817–818.

108. Ibid., 819–820.

109. "Jail Sentences For Speeders," *The Automobile*, June 8, 1905, 707.

110. "Eighteen Months in Prison," *The Automobile*, August 17, 1905, 199.

111. Ibid.

112. "Severe Punishment to Break Up Joy Riding," *The Horseless Age*, December 8, 1909, 674.

113. "In Touch with Market," *The Automobile*, October 5, 1905, 370.

114. "Man, Horse, Automobile, and the Highways," *The Automobile*, January 9, 1908, 39.

115. "Dangerous Animals on the Streets," *The Automobile*, February 22, 1906, 434.

116. Ibid.

117. Ibid.

118. "Massachusetts Automobile Law," *The Automobile and Motor Review*, June 14, 1902, 20.

119. "New Street Traffic Ordinance for New York," *The Automobile*, January 31, 1903, 154. Also, "Massachusetts Automobile Law Signed by Governor Bates," *The Automobile*, July 4, 1903, 21; and "Two Auto Bills Pending in Michigan," *The Automobile*, January 21, 1905, 160.

120. "The Horseless Carriage," 393–394.
121. Ibid.
122. George Morrison, "The New Epoch and the Currency," *The North American Review*, February 1897, 146.
123. Duryea, "As It Was in the Beginning," 47.
124. Ibid.
125. Latour, in his essay on door closers, states, "When humans are displaced and deskilled, nonhumans have to be upgraded and reskilled." The lack of discipline and reliability among humans, specifically porters, has to be compensated for, or put into order using nonhuman instruments, such as the hinge-pin or the door-closer. See Bruno Latour (writing as Jim Johnson), "Mixing Humans and Nonhumans Together: The Sociology of a Door-Closer," *Social Problems* 35, no. 3 (1988): 301.
126. Moffett, "The Edge of the Future," 154.
127. C. E. Woods would perhaps be described by Law as a "heterogeneous engineer" who marshaled physical as well as social resources to help build an automotive industry. See John Law, "Technology and Heterogeneous Engineering: The Case of the Portuguese Expansion," in *The Social Construction of Technological Systems: New Directions in the Sociology and History of Technology*, ed. W. E. Bijker, T. P. Hughes, and T. J. Pinch (Cambridge, MA: MIT Press, 1987), 111–134.
128. Baker, "The Automobile in Common Use," 200.
129. Woods, *The Electric Automobile*, 3.
130. "Pleasure vs. Commercial Cars," *The Automobile*, April 12, 1906, 651.
131. Ibid.
132. "Futile Deductions from Horse Statistics," *The Automobile*, March 29, 1906, 580.
133. Clay McShane and Joel A. Tarr, "The Centrality of the Horse in the Nineteenth-Century American City," in *The Making of Urban America,* ed. Raymond A. Mohl (Wilmington, DE: Scholarly Resources Inc., 1997), 119.
134. "The Horseless Carriage," 393.
135. "Farm Horse Giving Way to Its Rival, The Auto," *The Automobile*, February 20, 1908, 246.
136. "Horse-Drawn Statistics that Indicate Auto's Growth," *The Automobile*, February 6, 1908, 189.
137. F. R. Hutton, "Technical Experience" [contribution to "Trade Outlook for the Coming Year"], *The Automobile*, January 14, 1905, 42.
138. Woods, *The Electric Automobile*, 1.

139. Ibid., 29.
140. J. M. Davis, "A Carriage Trade Viewpoint," Letter Box [letter to the editor], *The Automobile*, May 18, 1905, 623.
141. Ibid.
142. "Carriage Builders Consider Autos," *The Automobile*, October 12, 1905, 413.
143. Roger Silverstone, *Television and Everyday Life* (London: Routledge, 1994), 98.
144. Stanley M. Jepsen, *The Coach Horse: Servant with Style* (South Brunswick, NJ: A. S. Barnes and Co., Inc., 1977), 26.
145. Charles Philip Fox and Jean Van Dyke, *Horses in Harness* (Greendale, WI: Reiman Associates, 1987), 118.
146. James E. Homans, *Self-Propelled Vehicles: A Practical Treatise on the Theory, Construction, Operation, Care, and Management of All Forms of Automobiles* (New York: Theo. Audel & Company, 1905), 62.
147. "Survival of the Horseless Carriage," 154.
148. Ibid.
149. Marius C. Krarup, "Influence of Carriage Styles on the Construction of Automobiles," *The Automobile*, April 11, 1903, 397.
150. Ibid.
151. "The Selection of a Gasoline Automobile According to Price," 424–425.
152. Krarup, "Influence of Carriage Styles," 397.
153. Ibid., 397–398.
154. Silverstone, *Television and Everyday Life*, 174.
155. "The Selection of a Gasoline Automobile According to Price," 424.
156. Laird, "The Car Without a Single Weakness," 797.
157. *Life*, June 2, 1904, 1127.
158. *Collier's*, May 23, 1908, 27.
159. *Collier's*, 1907–1908.
160. *Collier's*, February 22, 1908, 18.
161. "The Poor Man's Automobile," 49.
162. *Collier's*, 1907–1908.
163. *Collier's*, 1908.
164. "Motor Trucks Cheaper than Horses," *The Horseless Age*, April 12, 1911, 632.
165. *The Horseless Age*, December 29, 1909, 35.
166. *The Horseless Age*, December 8, 1909, 7.
167. Ibid., 9.
168. *The Horseless Age*, October 26, 1910, 3.

169. Ibid.
170. *Life*, January 13, 1916, 85.
171. According to Frank Luther Mott, author of the well-respected *History of American Magazines 1885–1905*, the first consumer automobile advertising appeared at the turn of the century, roughly around 1903. However, several consumer advertisements for automobiles had in fact appeared by the late nineteenth century.
172. *The Century*, October 1898, 48.
173. *The Atlantic Monthly* [*Advertiser* section], June 1901, 40.
174. *Collier's*, September 21, 1901, 4.
175. *Horseless Age*, December 8, 1909, 7.
176. *Life*, September 25, 1913, 533.
177. *Life*, January 1, 1901, 9.
178. *Life*, January 6, 1910, 40.
179. "Boy Knows His Car," *The Automobile*, October 17, 1903, 412.
180. "Youthful Pacific Coast Driver," *The Automobile*, January 22, 1906, 425.
181. *Life*, March 12, 1903, 3.
182. *McClure's Magazine*, June 1904.
183. Advertising work during this period generally involved spinning "webs of association around a product." For more information on advertising practices, see Ohmann, *Selling Culture*, 100–106.
184. *Life*, August 17, 1905.
185. Harry M. Collins, *Artificial Experts: Social Knowledge and Intelligent Machines* (Cambridge, MA: MIT Press, 1990), 221.
186. Ibid.
187. Woods, *The Electric Automobile*, 16.
188. Automobile drivers supposedly could earn anywhere from $25 to $100 in 1907–1908, according to advertisements for the New York School of Automobile Engineers that appeared in *Collier's* and *Life*. The school guaranteed job placement after two months of home study. Many jokes were made regarding the competence of drivers at this time, typified by a *Life* illustration of an owner tossed out of the car in a nasty accident, demanding that his chauffeur tell him where he learned to drive, to which the driver replies, "In a correspondence school, sir." See *Life*, May 24, 1906, 663.
189. Baker, "The Automobile in Common Use," 202.
190. Ibid.
191. Towle, "The Best Car for the Novice?," 395.
192. Turner, "Some Experiences with Modern Motor-Cars," 638.

193. Ibid.
194. Reynold Wik, *Henry Ford and Grass-Roots America* (Ann Arbor: University of Michigan Press, 1972), 100.
195. *The Horseless Age*, December 15, 1909, 14.
196. *Collier's*, March 23, 1901, 24.
197. *Life*, May 16, 1907, 669.
198. *Collier's*, 1907–1909.
199. *Collier's*, January 16, 1909, 10.
200. *Life*, May and June 1905.
201. *Collier's*, September 5, 1908, 5.
202. *Collier's*, May 23, 1908, 27.
203. *Collier's*, October 26, 1907, 29.
204. Davis, "A Carriage Trade Viewpoint," 623.
205. *The Saturday Evening Post*, June 2, 1923, 43.
206. *Collier's*, April 28, 1900, 25.
207. *Collier's*, April 18, 1908, 4.
208. *Life*, June 24, 1909, back cover.
209. *Collier's*, February 22, 1908, 18.
210. *Collier's*, June 30, 1900, 19.
211. *Collier's*, May 23, 1908, 27.
212. *Collier's*, April 18, 1908, 4.
213. Women at the turn of the century became purchasers and managers of commodities. Manufacturers through advertisements sought to establish a personal relationship with consumers to replace the role of merchants. (Ohmann, *Selling Culture*, 75–77).
214. *Life*, June 2, 1904, 523.
215. *The Saturday Evening Post*, April 2, 1904.
216. *Life*, May 5, 1904, 399.
217. *The Saturday Evening Post*, March 25, 1899, 619.
218. *Life*, 1904–5; *McClure's*, 1904–5.
219. "Considers Women Most Careful Drivers," *The Automobile*, September 5, 1907, 332.
220. *Collier's*, April 28, 1900, 25.
221. Scharff, *Taking the Wheel*, 167.
222. See the 1948 photo in Scharff, *Taking the Wheel*, unnumbered picture insert.
223. "Auto Women Shoppers," *The Automobile*, May 3, 1906, 734.
224. Dumay, "The Locomotion of the Future," 22.
225. Scharff, *Taking the Wheel*, 13.

226. Ibid., 167.
227. Towle, "The Coming of the Automobile," 33.
228. *Collier's*, 1901, 24.
229. *Life*, April 9, 1903; May 5, 1904. The Kelly-Springfield Tire Company also used women in their advertisements, even though they were selling tires for a horse-and-buggy outfit (*Life*, September 3, 1903).
230. *Collier's*, 1908.
231. Phebe Westcott Humphreys, "An Automobile Vacation on $1.60 a Day," *The Ladies Home Journal*, July 1906, 27.
232. Mrs. A. Sherman Hitchcock, "Women as Drivers of Automobiles," *The Automobile*, April 19, 1906, 674.
233. "Auto Women Shoppers," 734.
234. A. R. Ramsay, "Four Women and an Auto," *The Automobile*, June 24, 1909, 1044.
235. Mrs. A. Sherman Hitchcock, "Woman at the Wheel," *The Automobile*, May 6, 1906, 753.
236. Scharff, *Taking the Wheel*, 46–47.
237. Ibid., 37.
238. "Considers Women Most Careful Drivers," 332.
239. Thomas J. Fay, "Trend in Design and Fashion" [subsection "Female Labor is Now Being Utilized"], *The Automobile*, December 30, 1909, 1148.
240. Scharff, *Taking the Wheel*, 13.
241. "Motor Car Law," *Motor*, June 1909, 40–45. Also Scharff, *Taking the Wheel*, 25.
242. "Says Autos Are Replacing Sleighs," *The Automobile*, February 13, 1908, 220.
243. "Motor Life Stirring in the Iron City," *The Automobile*, December 5, 1903, 589.
244. "Club for Women Motorists in England," *The Automobile*, October 31, 1903, 451 (cover).
245. *The Automobile*, January 31, 1903, 148.
246. *Life*, September 3, 1903.
247. *Life*, January 6, 1910, 32.
248. The lowest price category includes Orient ($400); Ford ($400–$500); Olds, Reo, Holsman and Northern ($650); Cadillac, Wolverine, Mitchell, Pierce-Racine, Randall ($750); Maxwell ($780); Wayne, Queen, Holsman ($800); while the *highest* price category consisted of Lozier ($6,500); Simplex ($6,750); Napier, Fiat ($7,000); Matheson, Berliet, Hotchkiss, Martini; Rochet-Schneider, Renault ($7,500); English Daimler ($8,000);

Zust ($8,500); Panhard ($10,000); Mercedes ($10,900); Fiat ($12,000); Rochet-Schneider ($11,000); Panhard ($12,300). See "Chief Characteristics of 1906 Models," 32–66.

249. *Life,* May 13, 1915, 8.

250. "Casual Cut-Outs," *Dress and Vanity Fair*, October 1913, 108.

251. Ibid.

252. Ibid.

253. M. Worth Colwell, "America's First Track Race," *The Horseless Age*, February 1, 1911, 273.

254. "Fox Hunting with an Automobile," *The Automobile*, November 30, 1905, 59 (cover).

255. Victor Hart, "Hunting by Automobile in England," *The Horseless Age*, February 28, 1912, 1.

256. *The Saturday Evening Post*, June 30, 1923, 65.

257. *The Automobile*, July 1902, 587.

258. *Collier's*, July 4, 1908, 6.

259. *Life*, February 4, 1909, 163.

260. *Life*, September 13, 1906, 281. Also, a similar illustration with the same caption appeared in *Life*, September 17, 1914, 497.

261. "An Automobile Aid to Good Roads," *The Automobile*, April 8, 1905, 455 (cover).

262. "Poor Lo Takes an Automobile Outing," *The Automobile*, May 18, 1905, 609 (cover).

263. *Life*, March 4, 1909, 287.

264. John S. McGroarty, "The Valley of Surprise," *The West Coast Magazine*, June 1911, 266–267.

265. McShane and Tarr, "The Centrality of the Horse," 106.

266. Currently, the company is called William T. Pritchard Inc. of Ithaca, NY. The business has operated for five generations (118 years as of 2008), beginning with James G. Pritchard in 1890.

267. "Horse Firm Takes Auto Agency," *The Automobile*, April 1, 1909, 557.

268. *McClure's Magazine* and *Life* advertisements, 1905.

269. "Automobiles Again Displace Horses," *The Horseless Age*, April 10, 1912, 657.

270. "Autos Sold Before the Horse Block," *The Automobile*, May 31, 1906, 879.

271. "The Private Automobile Stable," *The Automobile and Motor Review*, July 5, 1902, 1. Also, advertisements for Woods Motor Vehicle, July 7, 1900, 29; advertisements of the Herreshoff Cars, *Collier's*, January 16, 1909, 10.

272. Flink, *The Automobile Age*, 135–140. Also, see p. 442, where the author indexed horses in terms of liabilities only (Horse—liabilities of the) without a corresponding discussion of assets.
273. "Automobiles vs. Horse-drawn Vehicles," *The Automobile*, August 10, 1905, 172.
274. Dumay, "The Locomotion of the Future," 22.
275. Fay, "Trend in Design and Fashion," 1148.

CHAPTER 3

CASE TWO: "BREEDING" THE JEEP: THE CONCEPTUALIZATION AND DIFFUSION OF THE IRON WARHORSE

When the automobile began to appear in public streets at the turn of the century, the U.S. Army began to explore various military applications of the newfangled machine.[1] Despite many years of experimentation and numerous attempts to motorize,[2] more than forty years elapsed before the army transitioned from muscle to motor power on an organization-wide scale. The U.S. military continued to use the horse during the inter-war years despite its ineffectiveness in World War I. The U.S. Cavalry in particular fought fiercely for the horse well into the early 1940s despite significant pressures from high-level military offices to motorize. In an environment more centralized than a consumer market, why did motor-ization take decades to implement?

In this second case I examine how the cavalry, the group most adherent to muscle power in the military, became motorized. Similar to the early years of the horseless carriage in civilian society, the motorcar was peripheral to the cavalry world and, to some extent, to the army. The horse, on the other hand, was central to the army's culture.[3] To even mention motorcars, at one point during the 1920s, was considered sacrilegious. The cavalry considered the horse a weapon, not just a form of transport.[4] After years of resistance, what made the cavalry give up their beloved horse?

The Cavalry Journal, an internal military publication from 1920–1946, showed the many conflicting sentiments about the motorcar during the interwar period. With a circulation of about 1,500–2,000,[5] *The Cavalry Journal* provided a forum for various ranking army officers to voice their opinions on matters concerning the cavalry. Those who recognized the need to modernize suggested various ways to maintain the cavalry's equine tradition on a new platform. Others who opposed any form of motorization, including many high-ranking military officers, insisted on being resolute regardless of technological changes sweeping the outside world.

Articles against motorization argued that the motorcar lacked the battle-field qualifications of the horse. The centrality of the horse to the cavalry's fighting principles demanded absolute faith in the horse. At the same time, this loyalty motivated many discussions on "breeding" an iron warhorse. In an ironic turn of events, the effort to preserve the horse led to the conceptualization of a successor materializing in the form of the jeep. *The Cavalry Journal* provides strong evidence of the role the horse culture played in impeding—and eventually facilitating—motorization.

For U.S. Cavalry operational standards, this study relies on the 1914 *Cavalry Service Regulations* issued to all cavalry regiments for indoctrination and training purposes. The same regulations also appeared in the new drill and service regulations manual of 1916.[6] No other succeeding cavalry manuals of these kinds were widely circulated. The 1914 *Cavalry Service Regulations* manual was issued immediately after a reorganization campaign in the cavalry in which the detached troops of enlisted cavalrymen who volunteered as park rangers at Yellowstone in 1872 were discharged from the Department of the Interior.[7] Needing to incorporate these volunteers, the U.S. Cavalry required definitive standards for its operations. It is conceivable that the 1914 manual was intended as a reference guide for cavalry principles and practices. This study uses these manuals as reference guides for the cavalry's combat culture.

To understand the initial years of motorization in the military, I examined how American and British soldiers used jeeps on the battlefront. The fieldwork of Virginia Cowles immediately after World War II,[8] a historian who interviewed members of the Special Air Services founded and led by British army officer David Stirling, provides the major source for evidence of the jeep's use in cavalry-style raids. Cowles's work has served as the basis for many succeeding historical accounts about Stirling. For a more individual level of analysis of soldiers and their jeeps, the work of Ernie Pyle, a well-known reporter and the first to conduct interviews by "embedding" himself with the troops, is a major source, as well as Bill Mauldin's poignant images and characterizations of the soldier with his jeep. Other accounts by World War II reporters and soldiers, as well as governmental documents, manuals, newspaper reports, feature articles, films, political cartoons, and secondary sources are among the other materials used to capture how the American soldier transitioned from muscle to motor power.

General Overview

After World War I, the U.S. Congress passed the National Defense Act of 1920 for the purpose of modernizing the army. Congress wanted to investigate why the United States entered the war with a wholly inadequate supply of men, arms, ammunition, and other military equipment, not to mention significant delays in transporting military support to Europe.[9] This disorganization harked back to a similar predicament at the outbreak of the Civil War in April 1861, when the Union government found itself utterly unprepared.[10] The 1920 Defense Act sought to ensure military procurement and industrial preparedness in the event of another war;[11] thus official doctrines for motorization[12] and mechanization[13] were created.

During the postwar years of 1919 to 1921, the War Department conducted a battery of tests to determine the feasibility of replacing the horse. The German mechanized attack of World War I forced the army to reexamine its traditional approach to warfare in earnest. Brigadier General George Van Horn Moseley, commander of the 1st Cavalry Division, stated, "When the cowboy down here is herding cattle in a Ford, we must realize that the world has undergone a change."[14] The motorcar appears to have gained some military support from a policy standpoint at this time, but the horse maintained its paramount position in the cavalry.

With the passing of the 1920 Defense Act, tanks came under the supervision of the U.S. Infantry because the army believed they would protect soldiers against an attacking enemy.[15] The act essentially positioned the tank to assume some of the battlefield roles performed by the cavalry.[16] The cavalry served to protect other combat arms,[17] thus the alarm within cavalry ranks at the idea of tanks serving as its replacement.

The cavalry assisted other combat arms by virtue of its ability to maneuver widely and freely. It was in charge of pursuing retreating enemies as well as covering the retreat of its own forces. The cavalry charged the flank and rear of the enemy and was responsible for delaying the enemy's advance until the arrival of other arms.[18] Thus it was often situated at a distance from the main body of the army.

The cavalry was also responsible for reconnaissance work, for controlling strategic positions, and for executing all types of raids because these attacks required speed and mobility.[19] It was the arm assigned to assist the infantry in filling the gaps in a firing line.[20] The cavalry, in essence, was first in the line of fire.

Despite the cavalry's revered position in the military,[21] its value came under fire with the advent of modernized warfare. *The Cavalry Journal* in 1922 described the controversy raging within the cavalry at the time.[22] Some groups wanted to embrace motorization, others sought to block any type of change, while still others tried to reconcile the two extremes by incorporating some level of motorization into the cavalry. Disagreements within the cavalry caused such internal upheaval that a public-relations battle ensued within its ranks. Many high-ranking cavalry officers were involved; some sought to preserve the status quo at all costs, while others preferred to directly confront and manage change.

When General Douglas MacArthur, acting as the new Chief of Staff in 1931, attempted to motorize and mechanize, he met fervent resistance. For instance, the cavalry refused to adopt MacArthur's term "mechanized force," insisting instead on the term "mechanized cavalry," which emphasized the precedence of horses over machines. Under MacArthur's directives[23] mechanization became a responsibility of both the chief of infantry and the cavalry.[24] From 1931 to 1940 mechanization developed along two separate lines with two different doctrines, although the so-called infantry fast tank and the cavalry combat car were exactly the same vehicle.[25] How, then, was the coexistence of the motorcar and the horse managed under the cavalry leadership?

One study argued that although the army intellectually appreciated the capabilities of the motor truck, emotionally its faith and trust remained with the horse.[26] The study found that a "feeling" of faith and trust in the horse was apparent in archival records from many years,[27] but it failed to cite any examples of how these emotions manifested themselves in a highly militarized environment. In the first sections of this chapter I attempt to uncover some of these sentiments and the driving force behind the cavalry's alliance with the horse.

WARHORSES AND CAVALRYMEN: INTERCHANGEABILITY IN IDENTITY

The bond between the cavalryman and his horse has been described as one of the oldest relationships in war.[28] Mounted attacks in the eighteenth century, initially carried out by soldiers who volunteered for the task, became a key strategic weapon of the U.S. Army in many wars during the nineteenth century.[29] Mounted rangers were appointed by Congress in 1832 to fight the American Indians on the western frontier, forming the first officially recognized cavalry in the regular army.[30] The use of horses influenced the pace and design of combat strategies; herein lies the beginning of the long-standing cavalry tradition in the U.S. Army.

The horse was intrinsic to battle and to the cavalryman's identity. A 1925 *Cavalry Journal* article described the relationship as one marked by deep loyalty: "Since civilization began the horse has been the companion, servant and friend of man."[31] More than ten years later, even with advances in automotive technology, the horse was still described as "far from being outmoded. Man's oldest servant of the animal creation is proving to be indispensable to the United States Army."[32]

The image of the cavalryman mounted on a horse charging toward the battlefield typifies a classic picture of a war hero. Without his horse, the cavalryman was just an infantryman. Indeed, cavalrymen were former infantrymen who traveled to the battlefield on horseback. Fighting at one time meant dismounting, finding cover, and shooting from the ground.[33] The horse provided the cavalry with speed and mobility. The emphasis on mounted attacks provided a psychological rallying point for soldiers to be aggressive and to take the offensive.

The 1914 *Cavalry Service Regulations* manual reiterated the need to "lean always toward the aggressive; to develop the 'habit of prompt decision'; and in cases of doubt, to take the boldest decision."[34] Ten years later, it was still a deeply held belief that "without mobility the cavalry [was] of little value."[35] Mobility allowed the cavalry to move across various terrains in search of enemies. It was believed that only horses could provide such capability.

The enemy and the terrain were the two elements critical to controlling the battlefield. According to the 1914 manual, "the character of the enemy and the nature of the terrain exercise controlling influence on the operations of the cavalry."[36] The soldier defeated the enemy through bold, offensive actions, while the horse conquered the terrain.

Dismounted actions were strongly discouraged because they hampered rapid, decisive movements. The 1914 manual stated that a "habitual reliance on dismounted action will weaken and eventually destroy initiative."[37] George S. Patton, an advocate of the mounted cavalry despite its wartime experience in the face of enemy tanks, believed that "success in war depends upon the golden rule [of] war. Speed—Simplicity—Boldness."[38]

Horses provided not only speed but also the necessary means for leaders to display boldness with sensational maneuvers and audacity on the battlefield. Leaders needed to be "active and energetic with a keen eye and good judgment, quick decision and a firm will," qualities considered essential to seizing control of the battlefield.[39] Failure to act decisively constituted a far "more serious charge against a cavalry officer" than acting too quickly, with a resulting mistake in operations.[40]

The use of horses allowed cavalrymen to be nimble and aggressive on the battlefield. Without horses, it was difficult for men to deliver surprise offensive attacks that would keep the enemy off-balance. The emphasis on offensive and prompt action was such that the 1926 War Department training regulations for the entire army insisted that only offensive attacks win wars and that every individual in the military must be "imbued with the spirit of the offensive."[41] This "spirit of the offensive" was deeply attached to the horse and the mobility it afforded.

The removal of the horse meant a loss of control, and even identity. It was believed that the fighting qualities of a good cavalryman could manifest themselves only through continued reliance on the horse. Without it, the cavalryman's fighting qualities would atrophy. Thus, while the cavalryman depended on the horse for survival and success during wartime, during peacetime he relied on the horse for the preservation of his fighting spirit.

> The directors of the Cavalry School believed that long periods of peace may, and often do, impair the efficiency of the cavalryman. There is a mental and a physical, as well [as] a mechanical, rut. Daring and aggressiveness, qualities that are indispensable to the mounted man, may easily be lost. They must be preserved at all odds. Only constant practice, day in and day out, can keep a man a bold and confident rider.[42]

Such was the dependency of the cavalryman on his horse: without it, he would not be ready for war. As late as 1940 a colonel confessed his attachment to his horse:

> The true cavalry soldier thinks of his horse's well being before his own—it is his friend and companion in danger; he enjoys going

> to war as long as he goes on a horse. It is a fact that a man while riding a blood horse, seldom had much fear of death.[43]

Motorization threatened the cavalry's way of life and style of warfare. Cavalrymen could not go into war without horses. Even after World War I, with the rise of trench warfare, poison gas, and powerful artillery, the cavalry remained resolute in its belief that mobility was the key to victory. According to a 1925 *Cavalry Journal* article, "Mobility, fire power, and shock are the characteristics of cavalry. The greatest of these is mobility."[44] What was different in 1925 was that mobility prepared the way for machine guns to execute a fatal attack,[45] whereas the 1914 *Cavalry Regulations Manual* recommended that cavalrymen themselves deliver the fatal blow, with machine guns serving only to clear their path.

While the 1914 cavalry-manual approach proved ineffective in World War I, the position of many high-ranking cavalry officers was that mobility remained indispensable to winning wars, regardless of who or what delivered the fatal blow to the enemy. As the Commander-in-Chief of the British forces in France stated after World War I, "The power of an army as a striking weapon depends on its mobility. Mobility is largely dependent on the suitability and fitness of animals for army work."[46] The horse was fervently held to be indispensable to the army's combat arsenal, despite its proven ineffectiveness against machine guns.

WARHORSES AND MEN OF WAR: INTERCHANGEABILITY IN FUNCTIONAL SPECIFICATIONS

To the cavalry, the horse provided the speed necessary to stay ahead of rapid changes on the battlefield. A major in the Ordnance Reserve described battles as incredibly rapid: "when things happen they happen very fast."[47] The enemy was a moving target, and a horse that could keep up was essential. Unlike in the civilian setting, where equine instinct provided many advantages in the carrying out of domestic duties, a warhorse had to be trained to suppress some of its instincts, particularly the need to flee from danger.

Horses had to follow cavalry commands, even if such commands led them charging toward death. Good warhorses possessed a high degree of courage, endurance, and speed, as well as an even temperament when faced with danger,[48] a description that could very well apply to a good cavalry leader. In this sense, the cavalryman's positive qualities were almost interchangeable with those of a good horse.

In order to be successful on the battlefield, a warhorse had to be trained, just like a soldier, to overcome its fears. Intelligence and instinct for self-preservation had to be suppressed in order to face the enemy. Warhorses would be "easily governed by the consistent application of recognized aids applied with kindness," and differences among warhorses "[would] be offset in part by intelligent training."[49] In many ways, these requirements describe a machinelike profile in which the horse's inherent instinct to flee from danger was suppressed so that it followed orders without resistance. There was also a machinelike quality to the military's "intelligent horse training," which differed from the civilian use of the term.

Intelligent horse training in the military would not be considered intelligent to the average civilian. Military horses had to be, first and foremost, easily governable and predictable. An army general described the importance of spending enough time in training a horse because "one great difficulty in raising new cavalry [was] that the horses [were] new, untrained, fractious, and that the men [had] not yet learned to ride."[50] While civilians would appreciate an intelligent horse that avoided danger, in the military, an intelligent horse would not exhibit independent thinking and would overcome its instinct for self-preservation at all costs.

Good warhorses, like good soldiers, needed to follow orders without resistance, and their peculiarities had to be ironed out, in a process much like standardization in mass production. What would emerge from the military's intelligent training was an army of fairly homogeneous warhorses with a range of predictable and controllable responses. Horses had to become "mechanical" in order to withstand the rigors of war.

The Percheron, a light draft horse, was considered ideal for military purposes because of its machinelike quality. Able to withstand the strain, exposure, and hardship of the battlefield, the horse could "best meet the exacting demands of modern warfare."[51] A light Percheron, weighing around 1,500 pounds, was considered the "Allies' most successful war-horse"[52] and was even described as "the best artillery horse the world had ever seen."[53] Indeed, the technical specifications of a mechanical horse would later prove to be nearly equivalent to the physical traits of the Percheron.

Horses were kept in reserve as long as possible in order to save their strength for battle. The U.S. Cavalry, emphasizing prompt action over planning, placed less emphasis on reconnaissance and hence delegated this task to motorcars. According to the 1914 cavalry manual, motorcars were to be used for reconnaissance to conserve the strength of the horse for combat.[54] The British fighting style, taken to be more deliberative, differed from the American style of boldness and prompt action.

The British cavalry valued reconnaissance highly and relied on the horse to perform this important task. The British cavalry commander typically dispatched the maximum possible number of cavalrymen to obtain strategic information for his plan of attack. The British valued planning and intelligence and used their best horses for investigative work, while the American forces regarded fighting power as more important and thus saved them primarily for the battlefield. Regardless of the approach, the horse was consistently used in those tasks deemed most critical.

The U.S. Cavalry's battle routines, combat style, and training revolved around the use of the horse. Faith and trust in the horse led the cavalry to assign strategic tasks to mounted rather than motorized troops. This trust in the horse could be understood as confidence in the familiar. Because cavalrymen understood horses better than motorcars, they presumably could better factor their weaknesses and strengths into their plans, and in this sense horses were more controllable and predictable in carrying out critical missions. At the same time, the cavalry's elite status also may have influenced the assignment of critical tasks.

While social-class hierarchies were not as prominent in the United States as they were in Great Britain, there was nevertheless a sense of superiority felt within the U.S. Cavalry with regard to other branches of the military.[55] The replacement of the horse meant the loss of the cavalry's elite status, along with its romantic imagery. The use of horses in Europe was also associated with nobility and aristocracy. During the Middle Ages, mounted knights were the highest-status warriors in England and elsewhere in Europe. The word "cavalry" comes from the French word *chevalier*, or mounted knight.

Thus the threatened removal of the horse constituted a foundational breakdown. A cavalry lieutenant-colonel wrote in the late 1890s that "without horses there can be no cavalry."[56] Twenty years later, a cavalry captain made a similar declaration: "Without the horse, the cavalry had been deemed 'practically useless.'"[57] According to a 1925 *Cavalry Journal* article, "The horse makes cavalry, distinguishes cavalry from foot troops, and gives the arm its characteristic of mobility. Cavalry therefore may operate within the powers of the horse and is held by his limitations."[58]

Another 1925 *Cavalry Journal* article may well have been a campaign to protect the paramount position of the horse in combat. The argument was that the horse was essentially indistinguishable from the cavalryman himself and from the entire cavalry as an organization. To further advance the cause of the cavalry, the author, a major general, argued that men and machines were dispensable but that horses were not: "There are many things in war that can not be done by men nor machines; and that only animals can do."[59] A soldier writing to his brother of his World War I experience advised, "The horses must come before anything else and you must always bear this in mind. No matter how much work is to be done on other jobs, be sure the horses have been attended first."[60] Such was the importance of horses that their welfare came first and foremost.

THE HORSE ON TRIAL: PRESSURES TO MOTORIZE

The wartime experience of the cavalry during World War I was painfully thin, however. Out of the seventeen cavalry regiments in the U.S. Army, only one, the 2nd Cavalry, was sent to Europe, and there it spent most of its time, as the last Chief of Cavalry Major General Herr described it, working on the "thankless and uncongenial task of running various remount stations."[61] Some U.S. cavalrymen sent to the battlefield in August 1918 to pursue retreating German soldiers were outfitted with such a "hodge-podge" of animals that the American riders resembled, again in the words of Major General Herr, "Don Quixote and Sancho Panza chasing after windmills."[62] The Chief of Cavalry felt that the respect and honor accorded to the heroes of the Civil War had dissipated. It was clear that the cavalry deeply resented the secondary role it had played in World War I. The once proud and brave cavalrymen were reduced to comic figures.

The fearlessness and prompt action that characterized the cavalry fighting style appeared foolish when wave upon wave of horses and men fell in front of German machine-gun fire. Despite lessons learned by the French and British cavalries early in the war, American soldiers fought in 1917 with the same aggressiveness and rashness inherent in the cavalry tradition. George Marshall explained the American disposition in 1918: "Our men gave better results when employed in a 'steamroller' operation, their morale suffered from delays under fire, their spirits were best maintained by continuous aggressive actions."[63] However, the U.S. Cavalry's offensive attacks near the end of World War I no longer commanded the admiration they had inspired in the American Civil War, because the nature of war had changed.

The "lightning strike" of German mechanized forces through Belgium during World War I rendered the U.S. Cavalry and the cavalry-style attack obsolete. Tanks replaced horse-drawn cannons; thus, despite the War Department's continued support of the U.S. Cavalry, it could no longer simply dismiss pressures to replace the horse. What, then, would justify the existence of the cavalry in the face of modern warfare? Although the U.S. Cavalry could not easily prove its importance abroad, it found a useful cause domestically—the capture of Pancho Villa. The services of the horse had been invaluable in guarding the long border with Mexico, where mountainous terrain and nonexistent roads rendered motorcars unusable.

Unfortunately for the cause of the horse, even in the Mexican expedition, the motorcar began to emerge as a viable alternative. Ironically, the very figure who actively sought to block motorization, George Patton, led the first publicized use of the motorcar in combat. In May 1916 then Lieutenant George Patton killed a critical member of Pancho Villa's guerrilla force, Julio Cardenas, and three other men, using the automobile as a form of warhorse. Patton and his men rode their automobiles much as cavalrymen used horses for mobility and mounted attacks. Patton later described this motorized warfare as simply a matter of employing the car like a horse and insisted that only cavalrymen were best suited to utilize the motorcar as such.[64] The combat application of the motorcar made headlines and popularized it as a potential combat weapon.

Suddenly cavalry combat was no longer *exclusively* associated with the horse. The stark contrast between the horse as a combat weapon and the car as a transport vehicle began to blur. Perhaps sensing this trend, Patton immediately sought to de-emphasize the glorious moment of the motorcar and insisted that the combat use of the automobile was simply a matter of the *continuation* of the great horse tradition.[65] But it was too late: the well-publicized shift from muscle to motor combat appeared in newspaper headlines, and the indispensability of the horse in military combat became less definitive. The motorcar's features, after all, were not radically different from what soldiers expected of a trained military horse. Horses had to be machinelike in facing the rigors of war.[66]

However, the cavalry starkly divided muscle and motor power. The War Department's higher-ranking officers, sympathetic to the long tradition of horse culture in the military, ruled against complete motorization in the 1920s by officially citing the motorcar's lack of cross-country capability as a significant limitation.[67] Modern warfare, many argued, was still a war of mobility. In a war of mobility, they saw the horse as irreplaceable.

In defining war by *where* it was fought rather than by *how*, one could argue that horses were the only sensible means by which soldiers could carry out their daily work. As Patton aptly phrased it,

> True, there are a limited number of gasoline neophytes who, while admitting the impossibility of using machines in such country, avoid the issue by the happy statement that, in future, wars will not take place in that sort of country. The futility of such evasions seems almost too flagrant to merit remark.[68]

The argument for the survival of the horse became a matter of discrediting the motorcar, the reverse of what happened with the horseless carriage in the first case study, in which the horse was discredited. The style of contention, however, was strikingly similar. In this case, it was the motorcar that failed to meet the physical standards set by the horse. Patton described how "obstacles that appear trifling to a well-mounted Cavalryman often put serious handicap upon machines."[69]

Motorcars could not function at night, in the fog, in storms, or in deep snow. Nor could they traverse mud and water as well as horses could. A cavalry captain stated categorically that "horses could explore portions of the terrain impracticable for motors."[70] Enemies would be alerted by the noise of an oncoming motorcar, which would also leave behind tire marks that could easily be traced. Patton argued that "without gasoline, machines are junk"; without this "priceless liquid," the mobility of soldiers would be seriously compromised, resulting in a situation "far more fatal than enemy fire."[71]

Horses, on the other hand, could eat anywhere and whenever possible. Furthermore, the motorcar lacked the wartime experience of the horse. For these reasons, the motorcar was *not the equivalent* of the horse. A major claimed that many wars had been fought without mechanical transport but that no wars had yet been fought without horses.[72] Another captain of the cavalry argued similarly: "The horse was first an animal of war, and it is inconceivable that war will ever be waged without him."[73] As late as 1939 the last Chief of Cavalry testified before a congressional committee that the horse had "stood the acid test of war," whereas the motorcar had not.[74]

Despite the German mechanized force attack in World War I, Patton, at that time a lieutenant colonel commanding the 304th Tank Brigade in 1918 in France,[75] stated, "A general survey of the tactical tendencies at the close of the World War seems to point to greater, and not lessened, usefulness and importance for cavalry."[76] The same argument was used by a cavalryman who insisted that modern warfare enhanced the cavalry's strength: "It has been said that the development of scientific and mechanical weapons renders the cavalry useless. On the contrary, these weapons give the cavalry an added efficiency."[77] These modern weapons, it was argued, could be easily incorporated into mounted attacks, although how such attacks could be carried out in mechanized warfare without significant casualties remained unclear.

Nevertheless, a captain stated, "War is a conflict between elements of flesh and blood, and inanimate armament is but a means by which it may be more successfully waged."[78] Fundamental to these arguments was the privileged position of the cavalry and the sanctity of the horse, regardless of technological changes sweeping the outside world. The cavalry toward the end of the 1920s became resolute, arguing a line of thought that demanded blind faith at a time when valid arguments had been advanced about the vulnerability of the horse in modern warfare.

An article attributed to "one of the faithful" stated, "As cavalrymen, we must have faith in the cavalry service, and we must have a doctrine which will allow other branches to see how well we keep the faith. To the cavalry itself, that faith must be sacred."[79] In cavalry terms, faith meant unquestioning acceptance of the status quo: its tradition, leadership, combat strategy, and, most of all, the inseparability of the horse from the cavalry.

This blind-faith position, in an organization that prided itself on having sentiment only upon the "execution of a plan drawn in cold reason,"[80] demanded that cavalrymen ignore rational arguments for motorization. A colonel stated it differently: "Every great general of the war has expressed his belief in the future brilliant role of cavalry. It is only the lesser individuals who refuse to be informed, close their eyes and choose to doubt."[81] Anyone who chose to critically examine the viability of the cavalry in modern war was described as ignorant and uninformed and was considered an outsider.

Adna Chaffee, often referred to as the "father of American armor," and who began his campaign for mechanization in 1927, had been told by the President of the War College to be "visionary and crazy" when he delivered a lecture calling for greater use of advances made in the automotive industry for military purposes.[82] Patton argued, however, that the incorporation of machines would only condemn the army to "disaster and defeat."[83] Again according to Patton,

> Regardless of the progress made in the development of fighting machines, Cavalry will always be necessary. It will hold its own because no other agency can perform Cavalry duties with equal reliability and dispatch. It can operate effectively in woods and mountains where machines cannot go; it can swim streams that would stop machines; and whether its supply trains come through or not, it can carry on day and night under any conditions of roads and weather. To expect mechanical vehicles—impotent without regular supplies, blind and deaf to control, and restricted by terrain—to take over these duties, is to expect the impossible.[84]

For Patton, there was no substitute for the horse. In an argument somewhat similar to that of Harry Collins on artificial intelligence, Patton pointed to the inability of the motorcar to match, item for item, each of the horse's capabilities. Thus Major General Williard A. Holbrook, Chief of Cavalry, recommended that if the cavalry were to motorize as required under the 1920 Defense Act,[85] it be for the purpose of enhancing the capabilities of horses, not supplanting them.[86] It is conceivable that Holbrook's approach was a concession for the purpose of *containing* the influence of the motorcar within the cavalry. The succeeding Chief of Cavalry, Major General Herbert B. Crosby, also used the same tack, suggesting in 1928 that machines should be used to supplement horses.[87]

However, the initial reconciliatory move of the Chiefs of Cavalry to incorporate motorization came to a halt when General Douglas MacArthur, as Chief of Staff in 1931, sought to institute an aggressive campaign to modernize the army.[88] In a written statement released by the War Department on May 18, 1931, he stated,

> Thus there has grown up in the public mind a very natural
> conception that Cavalry must include the horse. Modern firearms
> have eliminated the horse as a weapon, and as a means of trans-
> portation he has become, next to the dismounted man, the slowest
> means of transportation.[89]

MacArthur's mandate was in line with the growing trend toward motorization, which began with President Woodrow Wilson's comments about
war. Wilson, quoted by Major General George Van Horn Moseley in
a *Cavalry Journal* article, hinted at the obsolete style of boldness and
bravery in cavalry battles. "Modern wars are not won by mere numbers. They are not won by mere enthusiasm. They are not won by mere
national spirit. They are won by the scientific conduct of war, the scientific application of industrial forces."[90] This emphasis on "scientific
application" over the traditional values of patriotism and courage was
a direct hit to the idealized but nonetheless fundamental values of the
cavalry's fighting principle and the cavalryman's identity as a soldier.

One of the initial but symbolic moves to modernize the cavalry in the
1930s was to de-emphasize the importance of the horse in war and to
emphasize the role of modern weaponry. For instance, the pack the horse
carried containing firearms and ammunition became more vital than the
horse. A colonel noted that the pack was the most important element in
a cavalier's outfit:

> It requires no study to see at once that the pack element of the
> command is of the greatest importance in all cavalry operations.
> It includes the bulk of the fire power with its ammunition, the sig-
> nal communications, demolitions and messing facilities. Without
> its packs Cavalry would be reduced to approximately the power
> it had at the end of the Civil War and would be out of place in
> modern combat.[91]

The pack, a lifeless material, was now more important than the horse.
Some officers even believed that without the machine gun in the pack,
the horse would be rendered useless on the battlefield.[92] While the 1914
cavalry manual's fighting philosophy—an attitude that held throughout

much of the 1920s—placed modern weapons in a supportive role for mounted troops, the new emphasis on modernization in the 1930s placed firepower at the center of its combat strategy. Mounted troops came to be only as important as the packs they carried.

Horses, in this view, were no longer as critical as the mobility they provided. The cavalry kept its integrity as a combat arm only insofar as it delivered the appropriate amount of firepower. Indeed, a brigadier general stated in 1940, "With the increasing efficiency of hand firearms and their consequent increasing use, the horse soldier more and more found that his usefulness on the battlefield was limited."[93] Thus it appeared that in order to survive, the cavalry had to begin seriously considering military life without the horse. A lieutenant colonel observed,

> Our Cavalry is instinctively hostile to any machine which may supplant the horse, and inclined to disparage its effect. We are retreating to mountain trails and thick woods, hoping that no fast tank can follow. Our policy, on the contrary, should be to encourage the new arm, experiment with it, and bring out its characteristics, both favorable and unfavorable, since the place of the new arm in the army team, its missions and tactics, are far closer to those of Cavalry than they are to any other arm. The cavalryman is best able to understand its potentialities. It is improbable that a machine will ever be invented that is more efficient for all military purposes than the horse. But whether our cavalry divisions are completely mechanized or not, cavalry missions and cavalry tactics will remain, and the mechanized force will act in conjunction with the Cavalry.[94]

The author of this piece recognized the strengths and limitations of machines in serving military needs but nevertheless promoted their use. Like MacArthur, he sought to redefine the cavalry in terms of its mission and tactics rather than the means by which it went to war. As a cavalry major observed,

> We are living in a machine age and to be modern, Cavalry must take every advantage of the machines this age places at its disposal.

> This the Cavalry has done, and will do, more and more as these
> mechanical auxiliaries are developed and proved of value.[95]

Under MacArthur, the 1st Cavalry, the oldest and most respected regiment, was ordered to dismount and to begin experimenting with motorization and mechanization. This mandate was taken as a signal of the expansion of mechanical auxiliaries in the cavalry. With the move to disengage the horse from the cavalry, a search for what would now define the cavalry began.

Major General Guy V. Henry, Chief of Cavalry in 1932, articulated the crisis in the cavalry's organizational identity, asking, "If the cavalry is less important today than it has been in the past, of what are we speaking?" "No one knows," he said, "for there is no standardized conception of cavalry."[96] In opening up the definition of cavalry to something non-definitive, Henry disengaged the cavalry's traditional identification with the horse and launched a search for its new identity.

Cavalries in other countries faced similar pressures. For example, Germany also sought to define "what modern cavalry looks like, what it does and what can be required of it."[97] In more pessimistic terms, a French cavalier asked, "Will there be room for cavalry missions in the war of tomorrow?"[98] Italy's policy after World War I had also been to move away from using horses.[99] The proud tradition of the cavalry and its elite status shined less brilliantly without the horse.

THE "BREEDING" OF A CAVALRY CAR: CONCEPTUAL ORIGINS

The Chief of Cavalry, Guy Henry, was asked point-blank in a 1937 congressional hearing, "Is the horse obsolete?" Instead of answering the question directly, Henry responded with another question, "Can the modern iron horse supplant the animal horse?"[100] Henry reframed motorization *not* as the phasing out of the horse but as the transfer of capabilities from a biological to an iron form. The cavalry, despite its continued display of defiance, recognized the need to find a means to preserve equine tradition in the midst of an increasingly mechanized world. Similar to the discourse regarding motor power mimicking muscle power in the civilian setting, military conversations about motorcars articulated in equine terms appear to have opened doors to motorization.

As early as the 1920s, an ex-cavalryman writing in praise of the horse made a parallel between the circulating water system of the motorcar and the bloodstream of the horse. He compared the motor's radiator to the horse's lungs:

> Let us make a comparison between the horse and the motor. When the motor is working, heat is rapidly developed. When the temperature exceeds a certain degree, the efficiency of the motor is lessened, if it does not cease to function altogether. To accomplish the liberation of heat, most motors are provided with a circulating water system and a radiator. The water circulating through the heated working parts takes up the heat, or a portion of it, and carries it to the radiator, where it is liberated. The radiator is so constructed as to expose a maximum surface to the air. If this surface be appreciably reduced, the cooling out is retarded. In the horse, the blood stream and the lungs can be compared roughly to the water system and the radiator of the motor. The blood circulating through the working parts—the muscles—takes up toxins and heat and carries them to the lungs, where they are given off through expiration.[101]

This simple one-to-one correspondence between muscle and motor mechanics shows conceptual associations occurring around the time the cavalry was most adamantly against motorization. The description of the motorcar's cooling system, phrased in terms of the horse's biological workings, provides a conceptual link between mechanical and biological functions. The horse was described as releasing heat to the air through expiration, while the motorcar used the surface of the radiator to accomplish the same purpose. Although the cavalry saw the motorcar as operationally inferior to the horse, it was no longer sacrilegious to start making comparisons.

Another discourse regarding the motorcar, again phrased in terms of the superiority of the horse, relates to the need to "breed" an "iron horse" specifically designed for battle.[102] A colonel described how the cavalry should delineate the requirements needed for combat cars rather than make do with commercially designed vehicles:

> Therefore, in this day of the "Iron Horses" for use on cavalry missions we should fix our types and demand breeding (i.e. speed, correct design, and equipment) and insist upon these requisites. To adopt a policy for taking any cheap ill-bred scrub iron horse obtainable in quantity on the streets of the nation is certain to greatly reduce the effectiveness of mechanized cavalry units in the performance of their missions. It is a makeshift poorly planned procedure.[103]

The colonel sought to apply fundamental principles of horse breeding to the design of an ideal cavalry car. He rejected the idea of merely modifying commercial cars for cavalry use: "The cavalry knows by practical tests that such commercial vehicles do not meet all cavalry needs by any means."[104] Modified commercial cars, he argued, were "fundamentally unsound," and their supporters "display[ed] an ignorance of cavalry experimentation to date."[105] Despite derision about the commercial motorcar, conceptual links between muscle and motor were being established. More important, discussions about an ideal cavalry car began.

Many military authorities indicated that commercial cars, particularly the Ford Model T, were too light and fragile for combat.[106] On the other hand, large, ponderous vehicles, such as those used in World War I from the Four Wheel Drive Auto Company of Clintonville, Wisconsin, stood out, making an easy target for enemy fire.[107] The four-wheel-drive American Minneapolis-Moline truck was sturdy but sank in soft ground.[108] The standard Chevrolet two-wheel-drive passenger car, which had been tested by the military since the 1920s, was lightweight but did not perform well in cross-country terrain.[109] Marmon-Harrington's 1930s four-wheel-drive truck was similarly ineffective because it was too heavy for reconnaissance missions and too hefty for quick maneuvers on the battlefield.[110]

The problem with these commercial vehicles was that one desirable feature conflicted with another. The power, performance, and durability of four-wheel drive, for instance, conflicted with the need for light weight and agility on the battlefield. A cavalry major summarized the dilemma: "It is the combination of all desirable features which presents a problem."[111] He recounted that many cavalry officers demanded a sturdy machine that could traverse gullies, stumps, rocks, and carry various items of equipment and artillery, while at the same time they "[didn't] want one of these big, expensive monstrosities such as the present scout or combat car."[112] In other words, they wanted something that performed like what they were used to—the horse.

Patton expressed the same sentiments: "But, to be useful in any of the above capacities, the car must be mobile, practical, and simple to repair—not a costly, hypothetical monstrosity."[113] Patton was describing the light tanks the cavalry acquired from the infantry, which had been solely responsible for them as a result of the 1920 Defense Act.[114] When the cavalry assumed responsibility for light tanks in 1931 under the directives of Chief of Staff General Douglas MacArthur, they renamed them "combat cars" to distinguish their tanks from those of the infantry, but essentially they were the same tanks[115] (Figure 46).

In reference to the specification of robustness, speed, and mobility that these combat cars could not deliver, Major Grow stated, "Decisions must be made as to the relative importance of several contradictory factors."[116] He chose mobility as a top priority. In this regard, the horse remained unrivaled, particularly in cross-country mobility. Commercial vehicles could not fully satisfy this requirement because of the difficulty of combining light weight with durability. The horse, on the other hand, provided significant strength and power relative to its weight.

One can argue that much of the opposition to the motorcar at this time rested on its lack of combat readiness relative to the horse. The fastest car was not ideal because speed would compromise its sturdiness. A sturdy vehicle, on the other hand, would weigh more, compromising

FIGURE 46. Early cavalry cars.

Source. Colonel Bruce Palmer, "Mechanized Cavalry in the Second Army Maneuvers," *The Cavalry Journal* 45, no. 6 (November–December 1936): 462. Reprinted with permission of the U.S. Cavalry Association.

Note. Caption reads "COMMANDER OF THE MECHANIZED FORCE AND SOME MEMBERS OF HIS STAFF. Right to left: Colonels Bruce Palmer and Henry W. Baird, Lieutenant Colonels Willis D. Crittenberger and Guy W. Chipman, all 1st Cavalry; and Lieutenant Colonel Alvan C. Sandeford, 68th Field Artillery. Vehicles: Right, Colonel Palmer's command car; center, an armored car; left, artillery commander's car."

speed and agility on the battlefield. Deciding at what level of weight, power, and performance these specifications should be set resulted in much discussion in *The Cavalry Journal.*

TECHNICAL SPECIFICATIONS: THE IRON HORSE IN CONCEPTUAL FORM

Modified commercial vehicles failed to satisfy the needs of the cavalry. Many articles were written in the 1930s on the technical specifications of an ideal cavalry car. It is interesting that these articles consistently reiterated the same message: the cavalry car had to be lightweight, rugged, and have good visibility with cross-country capability—very much the description of a warhorse. The demand was not for the fastest, strongest, nor the most heavily armed vehicle.

As early as 1930 a lieutenant colonel began the dialogue in *The Cavalry Journal* on *possibly* entertaining the idea of replacing the horse:

> The development of motor driven vehicles has progressed far enough to make it possible for us, without undue strain on the imagination, to visualize a machine capable of maneuver and attack across nearly all types of terrain at a speed of from ten to sixty miles an hour.[117]

First Lieutenant H. G. Hamilton, Cavalry-Reserve, provided a comprehensive list of specifications in a 1935 article:

> It would appear that to fulfill the requirements of Cavalry, any type of machine used for reconnaissance should have extreme mobility, be of rugged yet light construction, possess an extended cruising radius, and be able to operate over cross-country terrain as well as on established roads.[118]

The three general specifications Lieutenant Hamilton listed capture the essence of what would later become known as the jeep: (1) it had to be inexpensive and easily produced so that in an emergency, such as a state of war, an ordinary commercial chassis could be easily converted into a wartime vehicle; (2) it had to be designed as a reconnaissance car, whose defensive measures were speed and concealment rather than heavy armor; and, finally, (3) it had to be able to traverse cross-country terrain at uniform speed.[119] In summary, it had to be "speedy, easily concealed amid trees, brush, or in small ravines, and capable of carrying a crew of from one to four men, depending on the mission."[120]

Major Grow described a vehicle of a "cavalry nature" similar to Hamilton's but sought to explain the inseparability of mobility from the cavalry's identity. "Mobility is the paramount characteristic for cavalry. Mobility alone cannot win the battle but without mobility our unit would cease to be a cavalry."[121] This statement could easily have come from the 1914 *Cavalry Service Regulations* except that the horse would now be mechanized. Grow argued that protection, while important, would not be as critical as speed, because speed would allow the necessary offensive attacks to destroy the enemy first.[122]

> Cavalry operations are comparable to open field running in foot-
> ball. Tacklers are likely to spring up from anywhere, flank or rear.
> Personnel in moving vehicles are rather helpless against surprise
> fire. Unlike the man on foot, they cannot drop to the ground
> instantly. They cannot even take rapid advantage of local cover
> usually available to the mounted man...Protection, however, must
> never predominate over mobility and fire power if we expect to
> carry out cavalry missions. Our real protection consists in striking
> the enemy first. Our objective must be to "Hit the other fellow
> before he hits you."[123]

Indeed, the bold offensive attacks prescribed in the 1914 manual, in
which protection meant being elusive rather than wearing heavy armor,
began to make sense. As Patton indicated, "Men who fought in tanks
would willingly dispense with 50 percent of protection in order to gain
5 percent of mobility."[124]

A colonel described the ideal cavalry car as having technical specifications similar to those Hamilton described: "cars should be lightweight, cross-country type with low visibility to the enemy—but with
visibility open to the sky for the crew, and minimum armor."[125] Another
cavalry general phrased it differently, but with the same emphasis on
the importance of balancing mobility, weight, fighting power, protection, and stability.[126] Too much weight meant sacrificing mobility. As
Patton—buttressed by his wartime experience with tanks—observed,
every ounce of extra weight would greatly reduce the machine's fighting
capability.[127] Thus, virtually all articles written on the ideal cavalry car
insisted on these two features—mobility and light weight.

The consistent emphasis on the importance of mobility could be
found in the many objections to the motorcar's dissimilarity to the horse.
What disqualified the machine was precisely the shining feature of the
horse—its ability to cross various types of terrain:

> Vehicular reconnaissance alone will never be able to give the
> commander all the definite information required. It must be
> gathered either by men on horseback or men on foot and, consid-
> ering varied terrain and the time element, horsed cavalry must be
> available.[128]

The arguments for the use of the horse and the technical specifications for the ideal cavalry car mirror each other. The emphasis on cross-country capability and the light weight requirements fit the description of a horse. The horse could easily be manhandled in various environments and was, at the same time, sturdy enough to withstand the pressures of war. It needed to be able to gear up immediately for high speed but also to be robust enough to traverse rough terrain. These were precisely the requirements converted commercial cars purportedly failed to meet. Thus the failure of such vehicles to serve military needs became the justification for keeping the horse.

In creating this ideal cavalry car, the War Department assigned the task of specifying the technical requirements to the offices of the Chief of Cavalry, the Cavalry School, the Cavalry Board, and the First Cavalry (Mechanized). The Chief of Ordnance was also assigned to the task, and it was the only non-cavalry office on the team. The Ordnance Department was experienced in utilizing the services of civilians in scientific matters and maintained close relations with industries during the interwar years.[129] These offices were ordered to devote their "utmost thought and attention" to crafting a car suitable for combat.[130]

THE BIRTH OF THE WARHORSE

The Chief of Cavalry, together with the Chief of Infantry, submitted to the Secretary of War on July 2, 1940, the technical requirements for a 4×4 quarter-ton prototype: maximum weight of 1,275 pounds, maximum wheelbase of 80 inches, and overall height of 40 inches.[131] By engineering standards, the weight requirement made no mechanical sense, according to Karl Probst, the engineer who designed the prototype. By the time all the military options were added, the weight of the vehicle would reach 2,000 pounds.[132]

Although no direct written evidence could be found which explicitly states that the military used the actual physical attributes of the Percheron as a template, there were nonetheless numerous articles in *The Cavalry*

Journal indicating the intent to create the mechanical equivalent of a war-horse. The striking resemblance between the technical specifications of the prototype and the physical attributes of the horse brings to mind the discourses on the "breeding" of a motorcar and the urgent need to preserve military life built around the horse.

The weight limit of 1,275 pounds,[133] for instance, roughly equals the average weight of a light Percheron.[134] The 40-inch height of the prototype is about the average height of shrubberies found in and around the eastern United States, which means that the prototype was intended to hide behind bushes to avoid enemy fire.[135] With the windshield up, the height of the prototype is 64 inches, or 16 hands in equine terms, equivalent to the height of an average Percheron. The 80-inch wheelbase, the distance between the front and the back wheel, is the average body length of a Percheron.[136]

Even if it were a mere coincidence that the technical specifications of the iron horse closely map onto the physical characteristics of an animal horse, a homegrown military device developed in the late 1930s by Major Robert Howie and Master Sergeant Melvin C. Wiley provides evidence of the failure of an "unhorselike" concept. Despite satisfying the cavalry's list of functional requirements, the "Howie machine-gun carrier," also popularly known as the "Belly Flopper,"[137] never captured the soldiers' imagination. Because of the need to maintain a prone position while driving it (see Figure 47) and the jarring one felt as a result of its lack of suspension (hence the name), it failed to glide like a "snake in the grass," as its inventors claimed.[138] Although the Belly Flopper has been accorded equal importance with the jeep prototype,[139] it was never used on the battlefield.[140]

It may be that ingrained practices built around the horse failed to be carried out by an object envisioned as a different animal despite its fulfilling on paper the military's requirements. What the Belly Flopper did contribute, however, was its chassis, which came from the British Austin Seven automobile.[141] It was the technical expertise of its franchise, the American Bantam Car Company, that provided the prototype for what became widely known as the iron warhorse.

FIGURE 47. Belly flopper.

Source. T. Richards, *Military Jeeps, 1941–45* (Brooklands Road & Track Series. Bloom-field, NJ; Cobham, Surry: Portrayal Press; Distributed by Brooklands Book Distribution, 1985), 6. Reprinted from page 66 of "The Jeep," *Life,* July 20, 1942, 65–71.

FEMININE ORIGINS OF THE JEEP

The original idea for the Austin Seven was born of the need to create a vehicle designed for women just learning to drive.[142] Its adaptability, maneuverability, and ease of handling account for the emphasis on its light weight. The Austin Seven was launched at the time of the British Roads Act of 1921, a law that taxed every automobile based on its horse-power.[143] The Austin originally weighed just over 700 pounds, with an overall length of 104 inches, width of 40 inches, and maximum speed of 50 mph when fully loaded.[144] The small, durable Austin Seven could seat two adults in front and a few small children in back, or one adult sitting sideways,[145] although it could—and often did—carry four adults.[146] It was considered a remarkable engineering achievement and was a huge com-mercial success in England but failed miserably in the United States.

In a desperate move to save the American Bantam Car Company, the sole licensed manufacturer of the Austin Seven, Roy Evans, a well-respected automobile dealer, took ownership of the company in 1932

and faced an oversupply of 1,500 unfinished cars, with the entire production facility at a standstill.[147] After years of trying to revive the fledgling
company, Evans approached the military in 1939, which at that time was
experimenting with small, highly maneuverable utility cars such as the
Belly Flopper. He provided the Pennsylvania National Guard with three
Bantams for testing as scout cars, and the rest, as they say, is history.[148]

The Pennsylvania National Guard was impressed with the performance of the Austins, and the following year, the military invited various
manufacturers to bid on creating a prototype of a scout car made to military specifications. Of the 135 manufacturers the military invited, only
two replied: American Bantam and Willys-Overland Motors, Inc. The
American Bantam Car Company won the bid, promising to complete a
pilot model within the specified forty-nine-day period, a deadline large
manufacturers decided they could not meet. With the exception of the
weight requirement, Karl Probst of the American Bantam Car Company,
hired as an independent contractor, fulfilled all of the military's specifications and—with the help of Bantam's factory manager and Detroit's
auto-parts suppliers—delivered the prototype on time.[149]

Approximately one month later, the Bantam Reconnaissance Command, 40 horsepower, known as the BRC 40, passed the military's grueling prototype tests in Maryland, witnessed by major manufacturers
such as Ford, Chrysler, General Motors, and Willys-Overland, who took
notes on the Bantam's design. The Bantam was driven through mud two
feet deep, over hills and various other types of terrain (Figures 48, 49,
and 50), and a cavalry general even ordered two men to see if they could
lift it out of a ditch.[150] The weight requirement was added to ensure that
the prototype could be manhandled easily like a horse.

After the Bantam passed the required vehicle tests, the head of the
inspection team, Major Lawes, announced his verdict:

> I have driven every unit the services have purchased for the last
> twenty years. I can judge them in fifteen minutes. This vehicle is
> going to be absolutely outstanding. I believe this unit will make
> history.[151]

FIGURE **48.** Testing the Bantam.

Sources. (A, B). Karl K. Probst with Charles O. Probst, "One Summer in Butler—Bantam Builds the Jeep," *Automobile Quarterly* 14, no. 4 (1976): 434. Photos courtesy of *AQ/ Automobile Quarterly.*
Note. Caption reads "Bantam Jeep testing during 1941: being helped over the rim-rock, Fort Bliss [A]: stream fording [B]."

FIGURE **49.** The Bantam in the mud.

Source. Karl K. Probst with Charles O. Probst, "One Summer in Butler—Bantam Builds the Jeep," *Automobile Quarterly* 14, no. 4 (1976): 435. Photo courtesy of *AQ/Automobile Quarterly.*
Note. Caption reads "Bantam Jeep testing during 1941: a demonstration at Ft. Myers."

A lieutenant colonel described the reaction of the 6th Cavalry, which received eight Bantam cars for testing:

> Within the short span of the past four months, the entire Army has become Bantam conscious. The novelty of these small cars has captured the fancy of all the arms and bids fair to effect major changes in our ideas on cross-country mobility.[152]

FIGURE **50.** The Bantam over the hills.

Source. Karl K. Probst with Charles O. Probst, "One Summer in Butler—Bantam Builds the Jeep," *Automobile Quarterly* 14, no. 4 (1976): 437. Photo courtesy of *AQ/Automobile Quarterly.*

The Bantam prototype weighed 2,200 pounds, far more than the 1,275 pounds originally specified. Since the jeep could easily be manhandled and performed all tasks expected of a horse, it came to be known as the new "Iron Pony."[153]

Several months earlier, the commanding officer of the 6th Cavalry had personally subjected the Bantams to "every conceivable test. From sticking them in mud holes to towing a 37-mm. gun," the iron pony proved able to perform satisfactorily.[154] The 6th Cavalry later tested the eight Bantams more extensively, using the plan outlined by their commanding officer. They found that the Bantam, like the horse, could operate cross-country with a gun in tow.[155] A lieutenant described in detail the performance of the Bantams:

The field tests themselves consisted of putting the car through mud and water, over rough ground, through brush, up and down banks, and over every adverse type of ground a vehicle could be expected to traverse—and over a good many that no other type could traverse. Through these tests, it has been proved, beyond any reasonable doubt, that these little cars can and will do far more than even the manufacturers expected...It has also been noticed that in the case of the Bantam, more than in any other, its ability to get out of tough spots is predicated directly on the boldness and ability of the driver.[156]

Two goals measured by the test reflected what was foremost in the minds of the cavalry officers: the ability of the vehicle to traverse cross-country terrain and the importance of the driver's boldness in getting "out of tough spots." The emphasis on the driver's role in exhibiting boldness certainly harked back to Civil War days, when wars were deemed to depend on dramatic offensive maneuvers by mounted soldiers. With the use of the Bantam, mechanized war was no longer simply a matter of operating machines; boldness played a central role too, much like the principle of leading described in the 1914 cavalry manual.

The verdict of the 6th Cavalry was that there was "no doubt in the minds" of the members of the regiment, including the commanding officer, that the little Bantam fulfilled the requirements of an ideal cavalry car.[157] It was just a year earlier that the 6th Cavalry had been most adamant, a staff correspondent reported, about keeping the horse. The 6th Cavalry believed that "there is no medium which exists that excels the mounted trooper for cross country mobility under all conditions."[158] The Bantam field tests convinced the 6th Cavalry to change its hard-line position:

1. Reconnaissance. "The Bantam upon receiving fire, upon observing the enemy, or upon catching sight of a physical road block darts quickly under cover off the road."[159] This ability to hide from the enemy but to maintain visibility satisfied the requirements of the cavalry in conducting reconnaissance missions. As mentioned by

a colonel in the early 1930s, ideal cavalry cars should have "low visibility to the enemy—but with visibility open to the sky for the crew."[160] The Bantam tested well in these types of situations. It was also deemed light enough to move across weak bridges but sturdy enough to reconnoiter on back trails.[161]

2. Noise factor. The Bantam moved quickly and quietly.[162] One of the main contentions about early motorcars, including motorcycles, was that their noise alerted enemies. Bantam cars on the roads were described to be "practically without noise."[163]

3. Supply. The Bantam could "certainly prove to be a life-saver" in transporting supplies, rations, gasoline, machine guns, and ammunition over difficult terrain, according to the examiners.[164] It was able to carry radios and other communications equipment, such as signal lamps and flags.[165]

The cavalry also continued to modify and improve the Bantam to better suit its needs—for example, affixing iron rails over the headlights for protection—but the overall result was overwhelming acceptance of the vehicle by users.[166] The vehicle eventually produced in mass quantities by Willys and Ford did not radically differ from Bantam's prototype or, for that matter, from the original specifications issued by the military, except perhaps for its weight.

Based on the U.S. War Department's technical manual, *1/4-Ton 4 X 4 Truck (Willys-Overland Model MB and Ford Model GPW)*, the jeep that went into final production did not deviate significantly from its original specifications, apart from the increase in weight and a slight increase in height with the windshield down (Table 1).

DISPLACEMENT OF THE HORSE

The infrastructure for teaching horsemanship also began to be taken over by the new iron pony. The Cavalry School began offering motor classes soon after the Bantam passed the cavalry's rigorous field tests. A major observed, "The Cavalry School, heart of the Cavalry service, opened a

TABLE 1. Comparison of original and production specifications.

Vehicle Specifications	Original Specifications	Willys MB/Ford GPW[a]
Performance	Four-wheel drive	Four-wheel drive
Weight	1275 lb	2453 lb
Payload	600 lb	800 lb
Wheelbase	80 inches	80 inches
Height		
with top up	64 inches	69.75 inches
with top down	36 inches	52 inches
Ground clearance	6.5 inches	8.75 inches

[a] Data from *The Complete WW2 Military Jeep Manual* (Hong Kong: Brooklands Books, Ltd., n.d.)

motor school, where under the central control and with the ablest personnel obtainable, motor specialists [were] trained."[167] It would have been inconceivable in the 1920s for such an arrangement to occur. However, with the materialization of the jeep, the horse infrastructure began to modernize in the early 1940s: driver maintenance, preventive maintenance, repairs, overhaul, and rebuilding were taught.[168]

Using the facilities of the Cavalry School, automotive education became part of cavalry curriculum. Automotive maintenance and repair were taught alongside horsemanship. No longer was motor power segregated from muscle power, as was the case during the 1920s. The procurement process, unlike the 1930s decentralized approach, became the provenance of one entity, the Ordnance Department. More important, the work routines of horse maintenance and care in the cavalry continued with automotive maintenance and repair.

By 1943 the Cavalry School included a Department of Motors, which was divided into five sections: Department Headquarters, Supply Sections, Automotive Section, Special Classes Section, and Tank Section. The Headquarters and Supply sections taught supervision, administration, and supply, while the Automotive Section taught all matters concerning wheeled vehicles. The Special Classes Section conducted all motor instruction, primarily for officers. The Tank Section was responsible for teaching the mechanical functioning, operation, and maintenance of all cavalry-type vehicles; the Bantam was most likely covered under this section.[169] In addition, the Cavalry Replacement Training School (CRTC) trained mechanics specializing in repair and maintenance (Figures 51 and 52).

Training in the use of motorcars for combat also took place at Fort Knox, Kentucky,[170] a traditional training ground for cavalry horses. Basic formations of the mechanized cavalry, such as the line, column, and echelon, closely resembled those of the horsed cavalry. Techniques of scouting and patrolling also remained the same as those of their horsed counterparts. Each crewman was responsible for his own vehicle, in much the same way a traditionally equipped cavalryman was expected to

FIGURE 51. Military mechanics in training.

Source. "Training Mechanics at C.R.T.C.," *The Cavalry Journal* 52, no. 2 (March–April 1943): 76. Reprinted with permission of the U.S. Cavalry Association.

FIGURE 52. Bantams in the Cavalry School.

Source. "The Cavalry School of 1943," *The Cavalry Journal* 52, no. 1 (January–February, 1943): 89. Reprinted with permission of the U.S. Cavalry Association.

care for his own horse. The conventional infantryman, by contrast, was used to looking after himself only. Routines of maintenance and care in the infantry were nonexistent, which the cavalry never failed to point out whenever it sought to justify its qualifications to control mechanized forces.[171]

The repair and maintenance of motorcars was centralized under the Quartermaster Corps. The strict performance specifications of the Bantam necessitated standardization in supply parts for efficient maintenance and repairs,[172] similar to the way standardization of horse behaviors was effected through intelligent training.[173] During the postwar years, the Quartermaster Corps worked closely with the automotive industry, attending important transportation conferences to influence the development of designs and inventions.[174] The Quartermaster Corps was responsible for maintaining auto parts, although the Ordnance Department shared some of this responsibility. In terms of combat doctrines, tactics, scout techniques, patrolling, and basic formations, cavalry fundamentals remained unchanged. River crossings, for instance (Figure 53), were taught in horse and motor versions.

The rapid diffusion of the Bantam, soon to be called the jeep,[175] may be partly explained by the ease with which it initially shared, then took over, institutions and infrastructure built around the warhorse. Motorization no longer required dramatic change in the cavalry's establishment; it also did not require a redefinition of the cavalry's strategy and identity as a combat arm. The Bantam, lightweight with cross-country capability, provided functionalities similar to those of the horse, which earlier, commercial versions of the motorcar had been unable to replicate. With the Bantam it was at last possible, in the eyes of soldiers, to decouple the horse from the cavalry.

WARHORSE IN ACTION: THE JEEP IN WORLD WAR II

Indeed, the jeep in World War II was declared the replacement for the horse.[176] The jeep was described by army men as a "blitz buggy," which could "scramble over rough country better than a horse."[177] Many other

FIGURE 53. Integrating jeeps into the Cavalry School curriculum.

Source. Lieutenants W. S. McCauley and R. M. Vance, "R.O.T.C. Graduates at the Cavalry School," *The Cavalry Journal* 51, no. 5 (September–October 1942): 76. Reprinted with permission of the U.S. Cavalry Association.

commercial motorcars, such as the famous Model T, and even homegrown army machines such as the Belly Flopper had failed to captivate the army's imagination. What made the jeep compelling?

The jeep's popularity among soldiers in World War II can be partly attributed to the ease with which they transferred practices related to animal warhorses to this new iron form. The jeep quickly replaced the horse in the highly intensified combat environment. The new "warhorse" provided a platform on which the spirit of the cavalry could be carried out in modernized warfare. The GIs' use of the jeep showed that the mechanical horse was able to execute many of the daily tasks usually performed by the animal horse. This versatility made the jeep almost indispensable everywhere and in this sense rendered it inseparable from the soldier.

The traditional bond between cavalrymen and their horses manifested itself in the way GIs cared for their jeeps, that is, giving them names and personalities, and bestowing attention on them as if they were sentient beings. In many ways, the GI saw his jeep in a way that horseless-carriage manufacturers forty years earlier would have wanted their customers to believe possible—as a mechanical contraption that in the eyes of its driver was a living thing.

Indeed, a sense of familiarity developed: soldiers' knowledge of their jeeps resembled the level of familiarity typically found between cavalrymen and their horses. For instance, some cavalrymen would know just by looking at a horse whether it was lame and, if so, whether in the front or back leg. Similarly, soldiers learned to know their jeeps inside out, to make band-aid fixes and patch-up repairs in order to cure their jeeps of damage and mechanical problems. The speed with which soldiers became skilled in using jeeps in various theatres of war indicates a seamless transfer of preexisting practices from the horse culture.

JEEP RAID AS CAVALRY RAID

One of the early applications of the jeep occurred in North Africa with the British army officer David Stirling, whose conception of desert raiders followed the basic principles of cavalry warfare in which shock and mobility were the key elements of an attack. Stirling's jeep raids blew up pipelines, attacked airfields, destroyed transport and fuel tankers, and essentially caused mayhem for the German general Erwin Rommel, known as the "Desert Fox." As Rommel described the situation, "These Commandos, working from Kufra, and the Qattara depression, sometimes operated right up into Cyrenaica, where they caused considerable havoc and seriously disquieted the Italians."[178]

In a single week, Stirling's highly trained, twenty-one-man team destroyed sixty-one planes and at least thirty vehicles without suffering any significant casualties.[179] Stirling's jeep raid, though unconventional for the English army,[180] resembled a cavalry-style attack as described in the 1914 *Cavalry Service Regulations* manual:

> The object of a raid, in the general case, is to strike the hostile army in the arteries upon which it depends for the flow of ammunition, reenforcements, food and supplies of all kinds indispensable to its efficiency.[181]

The most spectacular example of this type of raid in Stirling's portfolio was conducted in July 1942. The target was the airfield in Sidi Haneish, located on the Egyptian coastline less than a hundred miles west of El Alamein. This airfield was Rommel's major staging area, where planes constantly arrived in and departed from North Africa. Thus, it was perpetually full of aircraft, including badly needed transport carriers.

JEEP-CAVALRY FORMATION

Eighteen jeeps took part in the Sidi Haneish raid. The jeeps formed a double-line column that advanced, guns firing outward, between two rows of planes. One column of jeeps would destroy one row of planes. Stirling would lead the group, driving at the tip of the advancing columns, with the navigator right behind him. Stirling diagrammed his jeep formation as depicted in Figure 54.

The "jeep raid" formation Stirling popularized bore a striking resemblance to the two-column habitual formation illustrated in the 1916 *Cavalry Drill Regulations* manual.[182] The typical double-column formation of a cavalry also placed its senior noncommissioned officer at the head, leading the two columns about three yards ahead of the group. A two-yard distance was typically maintained between two cavalry columns (Figure 55).[183]

According to Stirling's plan, the jeeps would be about five yards apart. All jeeps, except for that of the navigator, would be equipped with a driver and two gunners; each gunner would handle two Vickers guns, each of which was capable of firing a thousand rounds per minute.

FIGURE 54. Stirling's attack formation.

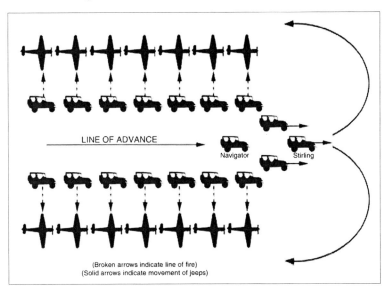

Source. Basic diagram from Virginia Cowles, *The Phantom Major: The Story of David Stirling and His Desert Command* (New York: Harper & Brothers Publishers, 1958). Modifications by Imes Chiu. Technical illustration by Jacqueline Côté-Sherman.

The first gunner sat in front with the two guns mounted before the driver, while the other gunner sat in the back, handling the other two guns. The jeeps maintained formation during the attack, circled around, and hit other planes on the outskirts of the field. Moments later, the jeeps disappeared into the dark desert, before the Germans could respond. Each jeep took a different path toward a common rendezvous point. The Sidi Haneish raid destroyed or damaged about 40 German aircraft and several tents and station buildings.[184] Stirling's spectacular jeep raid bore the qualities of a successful cavalry raid.

> The command must move rapidly and secretly. As a raiding force has no communications it is free to move in any direction. It should avoid serious combat except when necessary to accomplish

FIGURE 55. 1916 Two-column cavalry formation.

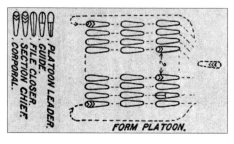

Source. Detail from *Cavalry Drill Regulations, 1916* (Washington, DC: GPO, 1917), 198.

its object or cut its way out. As far as possible, the enemy should be kept in ignorance of the position of the main body of the raiding force and of its destination and object of attack, and efforts should be made to deceive him as to future movements, especially as to the route of return.[185]

The similarity between the desert raiders' attack strategy and that of traditional cavalry warfare, including the striking resemblance of the jeep-raid formation to the cavalry's habitual two-column formation, shows that Stirling's approach was not original, although it has often been presented as such. The jeep certainly proved its cross-country capability in the trackless deserts of North Africa. Stirling's cavalry-style raids and combat strategy kept some things "unchanged" in the midst of modernization. The acceptance of the jeep might not have been as immediate and widespread had its use required fundamental changes in the soldier's tactics, routines, and mobility.

THE ALL-AMERICAN WONDER

The jeep was celebrated as "the most universally popular vehicle in the war; it was the one the Germans most liked to capture for day-to-day use."[186] General Dwight D. Eisenhower stated that senior officers considered the bulldozer, the jeep, the 2 1/2-ton truck, and the C-47 air-plane to be vital to their success in Africa and Europe.[187] General George C. Marshall considered the jeep the greatest contribution of the United States to modern warfare, while General Courtney Hodges described it as the "most useful motor vehicle we've ever had."[188] Ernie Pyle, the famous war correspondent embedded in the front lines, stated it differ-ently: "Good Lord, I do not think we could continue the war without the jeep. It does everything. It goes everywhere…The jeep is a divine instru-ment of wartime locomotion."[189]

The jeep's versatility was immediately recognized during the war, even by Axis troops, who were known to capture jeeps whenever pos-sible. The jeep's numerous applications included the following: ambu-lance to haul wounded men from the front lines, as it worked better than a regular ambulance because of its low silhouette; ground weapon, par-ticularly when mounted with machine guns, because of its maneuver-ability and mobility on the battlefield; weapons carrier; smoke-screen spreader, because of its ability to sprint across various terrain; plow for digging ditches to lay underground cables; tow truck to move smashed planes from bomb-pocked airfields; portable power plant for aircraft searchlights; mobile dump truck to fix airfields; field radio car; field telephone exchange; mobile antiaircraft unit; fuel and water supply car; fire-fighting unit; mobile air compressor; snow plow; railroad car when mounted with steel wheels, with the ability to pull as much as 52 tons of cargo; and a converted altar used by priests to conduct mass.[190]

The jeep's versatility endeared it even to those not on the front lines. Similar to the way horses had been used across a wide spectrum of social classes, the jeep was associated with people from all ranks and walks of life; from infantrymen to dignitaries such as presidents and prime min-isters like Franklin Roosevelt and Winston Churchill; royalty such as King George and Queen Elizabeth; high-ranking military officers, such as various commanders-in-chief and generals; and even Hollywood celebrities.[191] The fervent horse advocate General George S. Patton had a 1944 Ford Jeep custom-made to his liking.[192] One general who spent a significant amount of time on the front lines had been known to ride in a jeep for five to eight hours a day, insisting that, just like the Model T, the jeep was good for the liver.[193]

The "Four Jills in a Jeep" project, later turned into a film, further popularized the vehicle. The War Department commissioned four actresses to tour England, Ireland, and Africa in a jeep to perform for the troops.[194] General Marshall commented that this group of four entertainers was a huge success. Marshall suggested that in the future, perhaps two entertainers could be sent to front lines, and specifically mentioned transporting them to various ground units using "a single jeep."[195] In many ways, the jeep served as a social equalizer because no matter what station in life people came from, they liked to be seen riding in a jeep. However, the best-known companion of the jeep was the American GI.

THE GI AND THE JEEP

The motorized divisions of the U.S. Army in World War II were designed to be a hybrid of armored and infantry divisions, which, among other motorized vehicles such as tanks and modern weaponry, included jeeps.[196] Numerous stories about the inseparability of the GI from his jeep proliferated during the war. In a frequently cited and possibly apocryphal story related to the jeep, a watchman, usually identified as a Frenchman, was guarding his post at night when he suddenly heard a group of soldiers approaching on foot. He propped his machine gun to position and asked them to identify themselves. One of the members of the group replied that they were Americans. Without a moment's hesitation, the guard shot them all to death.

Later, the dead men were identified as German infiltration troops disguised in American uniforms. When asked how he knew they were Germans, the watchman replied triumphantly, "That's easy! Americans, they come in jeeps!"[197] This story has been recounted in many different versions in a wide range of popular books written about the jeep. The inseparability of the GI from his jeep was reminiscent of the inseparability of the cavalryman from his horse.

The jeep has been described as such a constant companion of the GI that the identity of the American soldier became inseparable from the jeep,[198] just as the cavalryman was never without his horse. This emotional attachment has been depicted in many different GI stories, particularly along the lines of soldiers crying over their shot-up jeeps and simply refusing to accept a replacement.[199] Similar to the way the horse became inextricable from the cavalryman and the cavalry as a whole, the jeep came to represent America and the American soldier. One of the classic images of World War II is the cartoon by Bill Mauldin of an old-time cavalryman shooting his broken-down jeep as one would put a dying horse out of its misery (Figure 56).

Another form of emotional expression came in the form of assurances, as a soldier confessed in World War II, that when new supplies of the "inevitable jeeps" would arrive together with other supplies, the soldiers felt more secure.[200] In Africa, jeeps brought new supplies of water and C-rations[201] every night.[202] Jeeps also brought the much-awaited mail from home.[203] GIs were so attached to their jeeps that they began to give them proper names. Bill Mauldin, himself a decorated war hero, referred to his jeep as a person with its own personality and idiosyncrasies:

> Two hundred miles is a long way for a jeep, even such a jeep as my pampered and well-manicured "Jeanie," who had covered more than ten thousand miles of Anzio, Italy, and France. The ordnance people called her the most neurotic jeep in Europe. But they cleaned out the carbon, ground the valves, and adjusted the carburetor. In spite of all this tender care, Jeanie developed ignition trouble on the way north and I had to stop every few miles in a pouring rain and get out and get under. After the first one hundred miles I was very glad that mud had obliterated the name "Jeanie" on the jeep's sides because I was swearing at the car in a way that would have crisped her namesake's lovely ears.[204]

Ernie Pyle wrote that it was customary throughout the army for soldiers to paint names on their vehicles. These were often girls' names, such as Suzy,[205] but some jeeps were called "Hitler's Menace" or "Invasion

FIGURE **56.** Cartoon.

Source. Copyright 1944 by Bill Mauldin. Courtesy of the Mauldin Estate.

Blues," and one was named "Bientot," which means "soon" in French.[206] Ernie Pyle observed that many soldiers might be ruthless fighters, but they were "just as sentimental as anybody else."[207]

Such was the intimacy shared by soldiers with their jeeps that, just as cavalrymen knew their horses well, GIs knew their jeeps inside and out and took good care of them. Ernie Pyle once recounted how a sergeant stopped a jeep in which Pyle was a passenger to ask the driver some questions. As he turned around to walk back to his own jeep, he ordered the driver to get his spare tire fixed, exclaiming, "Goddammit, why don't you take care of your vehicle?" "Spare tire?" replied the driver. "Yes, goddammit," the sergeant roared. "It's flat." While the sergeant was talking to the driver, he discovered the flat tire merely by feeling it with the slight pressure of his hand.[208]

The soldiers' intimate knowledge of their jeeps echoed the level of familiarity that was expected of cavalrymen with regard to their horses.[209] Many soldiers did take meticulous pride in their jeeps, and some kept them, in Ernie Pyle's words, "as neat as a pin."[210] As a film on the history of the jeep described it, "The average soldier forged a bond with their motorized horses."[211] The jeep design of the early models resembled that of a horse; it had no doors, no locks, and the ignition key was just a simple switch.[212] A soldier would hop into his jeep and drive away, just as one would mount a horse and take off.

The jeep was able to "scramble over rough country better than a horse. Cross streams. Climb rocky terrain. Beat down barbed wire entanglements. Dodge through forests. Hit more than 60 miles an hour on any solid road."[213] The jeep mirrored the description of an "iron horse" in the cavalry's imagination:

> "Iron horses" must be handy, able to turn and reverse in narrow lanes and other small spaces, and to negotiate obstacles; they must be capable of accelerating quickly, as well as moving fast, and must possess the quality of reliability.[214]

With the jeep, mobility as a combat strategy became associated with a mechanical medium. As the wartime jeep successfully assumed horse-related functions, a transfer to the vehicle of the physical and social infrastructure built around the horse occurred more smoothly—a transition that had not been achieved through direct mandates from high-level military commanders. Because its technical specifications were user-driven, the jeep carried out tasks usually performed by the animal horse; its versatility won the loyalty of the U.S. soldier, whose close bond with his jeep came to resemble the cavalryman's bond with "flesh and blood."

CONCLUSION

The jeep resurrected through mechanical means the spirit of the horse and the cavalry. Even in a hierarchical setting such as the military, the implementation of a new technological mechanism required persuasive measures that could not be achieved through authoritarian mandates alone. The jeep came to represent the final effort of the cavalry to preserve its cultural and military heritage. In trying to maintain the integrity of its organization after years of warding off pressures to motorize, the cavalry found a way to continue through mechanical means what had been its foundation—the horse.

Motorization framed in terms of equine sensibilities moved the motorcar in the 1920s into a more central role in the 1930s. Polarization within the cavalry with regard to motorization began to dissolve as meanings associated with the horse gradually became represented in the motorcar. The striking accuracy of many cavalry officers' descriptions of a successful replacement for the horse attests to the conceptualization of the jeep as something driven by practice rather than by technical considerations alone. The practices of cavalry-style leading, attacks on the enemy, and raids—essentially, the critical role of speed and mobility—drove discussions of an ideal car whose materialization came in the form of the jeep.

The jeep became the iron warhorse that allowed the cavalry to perpetuate its combat strategy of speed and mobility and its identity as the protector of other combat arms.[215] The cavalry specifically "bred" a motorcar designed to mimic the horse. Tested and proven to work like a horse, the jeep rapidly diffused, assuming existing infrastructures and relations built around the horse. The immediate diffusion of the jeep, largely attributed to its "breeding," occurred seamlessly as users transferred practices from an animal to an iron form.

The first case study showed how justification for a mechanical horse occurred largely through rhetoric and imagery. This second case study showed the actual realization of a mechanical horse. In both cases, organizational survival was at stake. Manufacturers had to win over horsemen to survive in the nascent automotive industry, whereas cavalrymen had to defend their role in an increasingly modernized world. The concept of the horse provided the means to maintain the old in a new mechanized body. The motorcar, a much-hated and controversial contraption, moved out of the periphery into the mainstream as a result of its association with the horse.

In the final case of this three-part study I examine the diffusion of the jeep in a new, localized form. After the war, the jeep as a World War II army-surplus vehicle in the Philippines evolved into a custom-built, elaborately ornamented passenger vehicle that came to be known as the jeepney. The jeepney eventually took over and expanded the Philippine transportation system, which originally relied on the horse.

ENDNOTES

1. Norman Miller Cary Jr., "The Use of the Motor Vehicle in the United States Army, 1899–1939," (PhD diss., University of Georgia, 1980), 5.
2. The War Department in 1928 defined motorization as the replacement of animal transport with motor vehicles (David E. Johnson, *Fast Tanks and Heavy Bombers: Innovation in the U.S. Army, 1917–1945*, Cornell Studies in Security Affairs [Ithaca, NY: Cornell University Press, 1998], 98).
3. Ibid., 125.
4. United States, War Department, Office of the Chief of Staff, *Cavalry Service Regulations, United States Army (experimental), 1914* [hereafter cited as *Cavalry Service Regulations, 1914*] (Washington, DC: GPO, 1914), 9.
5. *American Newspaper Annual and Directory* (Philadelphia: N. W. Ayer & Son, 1924, 1930, 1935). Also, *Union List of Serials in Libraries of the United States and Canada*, Third Edition, Volume 1, A–B, ed. Edna Titus Brown, under the sponsorship of the Joint Committee on the Union List of Serials, with the cooperation of the Library of Congress (New York: H. W. Wilson Co., 1965). The *Union List of Serials* traces the origins of *The Cavalry Journal* to *Armor*, which began publication in 1888. Then the journal suspended publication from January 1900 to June 1902, and then again from July 1918 to January 1920. From April 1920 to June 1946, *The Cavalry Journal* was published. Then from July/August 1946 through May/June 1950 the journal was published under the name "Armored Cavalry Journal" (*Union List of Serials*, 485). The circulation of *The Cavalry Journal* during its initial years of publication was approximately 2,000 (*American Newspaper Annual and Directory*, 158), but it seems to have gradually decreased through the years. In 1930, the circulation decreased to 1,600 (*American Newspaper Annual and Directory*, 156). In 1935, *The Cavalry Journal* was published bimonthly rather than quarterly, but the circulation again decreased to 1,458 (*American Newspaper Annual and Directory*, 140).
6. Mary Lee Stubbs, Stanley Russell Connor, and U.S. Dept. of the Army, Office of Military History, *Armor-Cavalry Part I*, Army Lineage Series (Washington, DC: GPO, 1969), 32–33; United States, War Department, Office of the Chief of Staff, *Cavalry Drill Regulations, 1916*, War Department Document No. 340 [hereafter cited as *Cavalry Drill Regulations, 1916*] (Washington, DC: GPO, 1917).

7. Stubbs and Connor, *Armor-Cavalry Part I*, 32–33.

8. Virginia Cowles, *The Phantom Major: The Story of David Stirling and His Desert Command* (New York: Harper, 1958).

9. Major-General William Crozier, *Ordnance and the World War: A Contribution to the History of American Preparedness* (New York: Charles Scribner's Sons, 1920), vii.

10. A. Howard Meneely, *The War Department, 1861: A Study in Mobilization and Administration*, Studies in History, Economics and Public Law, Vol. 300 (New York: Columbia University Press, 1928), 7.

11. Charles R. Shrader, *United States Army Logistics, 1775–1992, An Anthology*, CMH Pub., Vol. 68 (Washington, DC: Center of Military History, 1997), 457–464.

12. The War Department in 1931 issued an official definition of motorization and mechanization. Motorization is "the substitution of the motor-propelled vehicle for animal-drawn in the supply echelons of all branches of the Army, and in providing increased strategical mobility for units of all types through the carrying of men, animals and equipment in motor vehicles over roads." Mechanization is defined as "the application of mechanics directly to the combat soldier on the battlefield." See Captain Arthur Wilson, "The Mechanized Force: Its Organization and Present Equipment," *The Cavalry Journal* 40, no. 165 (May–June 1931): 7.

13. Mechanization is the use of motor vehicles for combat.

14. Brigadier General George Van Horn Moseley to Maj. Gen H. B. Crosby, December 9, 1927, File 322.02, OCC. Box 12, RG 177, NA, in Johnson, *Fast Tanks and Heavy Bombers*, 125.

15. *Encyclopedia of American Military History*, vol. 3, P to Z, gen. ed. Spencer C. Tucker, assoc. eds. David Coffey, John C. Fredriksen, and Justin D. Murphy (New York: Facts on File, 2003), 840.

16. Stubbs and Connor, *Armor-Cavalry Part I*, 49.

17. *Cavalry Service Regulations, 1914*, 9.

18. Ibid., 10.

19. Ibid.

20. Ibid.

21. Johnson, *Fast Tanks and Heavy Bombers*, 27.

22. Captain Gordon Gordon-Smith, "The Role Played by the Serbian Cavalry in the World War," *The Cavalry Journal* 31, no. 128 (July 1922): 245.

23. *Encyclopedia of American Military History*, vol. 3, P to Z, 840–841.

24. Johnson, *Fast Tanks and Heavy Bombers*, 117.

25. Ibid., 118. Also in Stubbs and Connor, *Armor-Cavalry Part I*, 52.

26. Cary, "The Use of the Motor Vehicle in the United States Army, 1899–1939," 83.
27. Ibid.
28. R. L. DiNardo, *Mechanized Juggernaut or Military Anachronism? Horses and the German Army of World War II*, Contributions in Military Studies, Vol. 113 (New York: Greenwood Press, 1991), 79.
29. Gregory J. W. Urwin, *The United States Cavalry: An Illustrated History, 1776–1944*, Red River Books ed. (Norman: University of Oklahoma Press, 1983), 9–54.
30. Major General William Harding Carter, "Early History of American Cavalry," *The Cavalry Journal* 34, no. 138 (January 1925): 7–8.
31. Wayne Dinsmore, "What Every Horseman Should Know," *The Cavalry Journal* 34, no. 140 (July 1925): 292.
32. "Horses and Motors," *The Cavalry Journal* 45, no. 194 (March–April 1936): 105.
33. Richard Wormser, *The Yellowlegs: The Story of the United States Cavalry*, 1st ed. (Garden City, NY: Doubleday and Company, Inc., 1966), ix.
34. *Cavalry Service Regulations, 1914*, 221–223.
35. Lieutenant Colonel Clarence Lininger, "Mobility, Fire Power, and Shock," *The Cavalry Journal* 34, no. 139 (April 1925): 178.
36. *Cavalry Service Regulations, 1914*, 222.
37. Ibid., 221.
38. Carlo D'Este, *Patton: A Genius for War*, 1st ed. (New York: HarperCollins Publishers, 1995), 306.
39. *Cavalry Service Regulations, 1914*, 223.
40. Ibid.
41. Lieutenant W. F. Pride, "Principle of the Offensive," *The Cavalry Journal* 35, no. 142 (January 1926): 55.
42. Oliver McKee, Jr., "With the 'Cavalree' at Fort Riley," *The Cavalry Journal* 34, no. 138 (January 1925): 72.
43. Colonel H. S. Stewart, "Mechanization and Motorization: The Final Chapter Has Not Been Written," *The Cavalry Journal* 49, no. 217 (January–February 1940): 41.
44. Lininger, "Mobility, Fire Power, and Shock," April 1925, 178.
45. Ibid., 178–179.
46. Captain Sidney Galtrey, *The Horse and the War* (London: Country Life, 1918), 11.
47. Major James R. Randolph, "Mental Mobility," in *The Cavalry and Armor Heritage Series*, ed. Royce R. Taylor, Jr., 59 (Fort Knox, KY: United States

Armor Association, 1986). Reprinted from *The Cavalry Journal* 49, no. 217 (January–February 1940): 10*.

48. An Ex-Cavalryman, "By Their Horses Ye Shall Know Them," *The Cavalry Journal* 33, no. 135 (April 1924): 199.

49. "Fundamentals of Cavalry Training Policy," ed. Jerome W. Howe, *The Cavalry Journal* 30, no. 123 (April 1921): 184.

50. Brigadier General James Parker, "The Cavalryman and the Rifle," *The Cavalry Journal* 37, no. 152 (July 1928): 366.

51. Galtrey, *The Horse and the War*, 124.

52. Ibid., 123.

53. Ibid., 128.

54. *Cavalry Service Regulations, 1914*, 227.

55. Stubbs and Connor, *Armor-Cavalry Part I*, 3.

56. Lieutenant-Colonel Jean Jacques Théophile Bonie, *The French Cavalry in 1870*, ed. Captain Arthur L. Wagner, trans. C. F. Thomson, in *Cavalry Studies from Two Great Wars*, International Series, vol. 2 (Kansas City, MO: Hudson-Kimberly Pub. Co., 1896), 125.

57. Gordon-Smith, "The Role Played by the Serbian Cavalry in the World War," 245.

58. Lininger, "Mobility, Fire Power, and Shock," 181.

59. Major General James G. Harbord, "The Part of the Horse and the Mule in the National Defense," *The Cavalry Journal* 35, no. 143 (April 1926), 159.

60. "The Horses Come Before Anything Else," *The Cavalry Journal* 37, no. 152 (July 1928): 415.

61. Major General John K. Herr and Edward S. Wallace, *The Story of the U.S. Cavalry 1775–1942*, 1st ed. (Boston: Little, Brown and Company, 1953), 243.

62. Ibid.

63. Urwin, *The United States Infantry*, 166.

64. Martin Blumenson and George S. Patton, *The Patton Papers* (Boston: Houghton Mifflin, 1972), 331–337.

65. Ibid., 337.

66. This example supports Dreyfus's claim that it is more feasible for humans to become like machines than vice versa. For details, see Hubert L. Dreyfus, *What Computers Can't Do: A Critique of Artificial Reason* (New York: Harper & Row, 1972).

67. Cary, "The Use of the Motor Vehicle in the United States Army," 133–136. See also Johnson, *Fast Tanks and Heavy Bombers*, 56.

68. Major George S. Patton, "Motorization and Mechanization in the Cavalry," *The Cavalry Journal* 39, no. 160 (July 1930): 333.

69. Major G. S. Patton, Jr., "Mechanization and Cavalry," *The Cavalry Journal* 39, no. 159 (April 1930): 236.

70. Captain Leonard Nason, "Horse and Machine," *The Cavalry Journal* 38, no. 155 (April 1929): 193.

71. Patton, "Motorization and Mechanization in the Cavalry," 345.

72. Harbord, "The Part of the Horse and the Mule in the National Defense," 159–160.

73. Captain George L. Caldwell, "A History of Cavalry Horses," *The Cavalry Journal* 37, no. 153 (October 1928): 557.

74. "Cavalry Affairs before Congress," *The Cavalry Journal* 48, no. 211 (January–February 1939): 132.

75. Stubbs and Connor, *Armor-Cavalry Part I*, 45.

76. Major George S. Patton, Jr., "What the War Did for Cavalry," *The Cavalry Journal* 31, no. 127 (April 1922): 169.

77. Kirby Walker, "Cavalry in the World War," *The Cavalry Journal* 33, no. 134 (January 1924): 11.

78. Caldwell, "A History of Cavalry Horses," October 1928, 557.

79. "One of the Faithful," "Faith in and a Doctrine for the Cavalry Service," *The Cavalry Journal* 36, no. 147 (April 1927): 227.

80. Major A. D. Surles, "Cavalry Now and to Come," *The Cavalry Journal* 40, no. 164 (March–April 1931): 5.

81. Colonel H. S. Hawkins, "The Importance of Modern Cavalry and Its Role as Affected by Developments in Airplane and Tank Warfare," *The Cavalry Journal* 35, no. 145 (October 1926): 489.

82. Colonel Wesley W. Yale, General I. D. White, General Hasso E. von Manteuffel, *Alternative to Armageddon* (New Brunswick: Rutgers University Press, 1970), 76–78.

83. Patton, "Motorization and Mechanization in the Cavalry," July 1930, 334.

84. Patton, "Mechanization and Cavalry," April 1930, 237.

85. George Vidmer, "Major General William Ames Holbrook," *The Cavalry Journal* 41, no. 174 (November–December 1932): 42.

86. Johnson, *Fast Tanks and Heavy Bombers*, 124–126.

87. Ibid., 126.

88. John B. Wilson, *Maneuver and Firepower: The Evolution of Divisions and Separate Brigades*, Army Lineage Series (Washington, DC: Center of Military History, United States Army, 1998), 123.

89. "Mechanized Force Becomes Cavalry," *The Cavalry Journal* 40, no. 165 (May–June 1931): 5.

90. Major General George Van Horn Moseley, "Industry and National Defense," *The Cavalry Journal* 40, no. 162 (January 1931): 18.

91. Colonel Daniel Van Voorhis, "Packs and Leading," *The Cavalry Journal* 39, no. 161 (October 1930): 498.

92. First Lieutenant Wesley W. Yale, "The Influence of Pack Loads on the Employment of Cavalry," *The Cavalry Journal* 43, no. 184 (July–August 1934): 19.

93. Brigadier General Henry J. Reilly, "Horse Cavalry and the Gas Engine's Children," *The Cavalry Journal* 49, no. 217 (January–February 1940): 3.

94. Lieutenant Colonel K. B. Edmunds, "Tactics of a Mechanized Force: A Prophecy," *The Cavalry Journal* 39, no. 159 (July 1930): 410.

95. Major E. C. McGuire, "Armored Cars in the Cavalry Maneuvers," *The Cavalry Journal* 39, no. 160 (July 1930): 397.

96. Major General Guy V. Henry, "The Trend of Organization and Equipment of Cavalry in the Principal World Powers and Its Probable Role in Wars of the Near Future," *The Cavalry Journal* 41, no. 170 (March–April 1932): 5.

97. Lieutenant General G. Brandt, "Why is the Cavalry Still Necessary?" *The Cavalry Journal* 41, no. 171 (May–June 1932): 46.

98. Chef D'Escadrons Breveté Mariot, "The Cavalry's Problem," *The Cavalry Journal* 43, no. 183 (May–June 1934): 14.

99. John Joseph Timothy Sweet, "Ferrea Mole, Ferreo Cuore: The Mechanization of the Italian Army, 1930–1940" (PhD diss., Kansas State University, 1976), 150.

100. "Necessity for Horsed Cavalry Under Modern Conditions: Extract from the Recent Hearings Before the Subcommittee of the Committee on Appropriations, House of Representatives, on the War Department Appropriation Bill, 1938," *The Cavalry Journal* 46, no. 201 (May–June 1937): 251.

101. An Ex-Cavalryman, "By Their Horses Ye Shall Know Them," 199–200.

102. Colonel Charles L. Scott, "Progress in Cavalry Mechanization: Scout Car Developments," *The Cavalry Journal* 45, no. 4 (July–August 1936): 281.

103. Ibid., 284.

104. Colonel H. S. Stewart, "Mechanized Cavalry Has Come to Stay," *The Cavalry Journal* (November–December 1938): 284.

105. Ibid.

106. Konrad F. Schreier, Jr. "The Military Model T Ford," *Military Collector & Historian* 39, no. 3 (1987): 99.

107. Arch Brown and the Editors of *Consumer Guide, Jeep: The Unstoppable Legend* (Lincolnwood, IL: Publications International, Ltd., 2001), 11–12.

108. Graham Scott, *Essential Military Jeep: Willys, Ford & Bantam Models, 1941–45* (Bideford, Devon, U.K.: Bay View Books Ltd., 1996), 7.

109. Konrad F. Schreier, Jr. "Born for Battle," *Military Jeeps 1941–1945* (Surrey: Brooklands Book Distribution Ltd., n.d.), 30.

110. Scott, *Essential Military Jeep*, 9.
111. Major Robert W. Grow, "Military Characteristics of Combat Vehicles," *The Cavalry Journal* 45, no. 6 (November–December 1936): 508.
112. Ibid., 509.
113. Major G. S. Patton, "Armored Cars with Cavalry," *The Cavalry Journal* 33, no. 134 (January 1924): 10.
114. *Encyclopedia of American Military History*, vol. 3, P to Z, 840–41.
115. Ibid. Also in Johnson, *Fast Tanks and Heavy Bombers*, 118. Also in Stubbs and Connor, *Armor-Cavalry Part I*, 52.
116. Grow, "Military Characteristics of Combat Vehicles," 509.
117. Edmunds, "Tactics of a Mechanized Force: A Prophecy," 410.
118. First Lieutenant H. G. Hamilton, "A Light Cross-Country Car," *The Cavalry Journal* 44, no. 189 (May–June 1935): 30.
119. Ibid. Also in Grow, "Military Characteristics of Combat Vehicles," 510.
120. Hamilton, "A Light Cross-Country Car," 30.
121. Grow, "Military Characteristics of Combat Vehicles," 509.
122. Ibid., 510–511.
123. Ibid., 509.
124. Patton, "Armored Cars with Cavalry," 10.
125. Colonel Albert E. Phillips, "The First Motorized Cavalry," *The Cavalry Journal* 43, no. 183 (May–June 1934): 10.
126. Grow, "Military Characteristics of Combat Vehicles," 508.
127. Blumenson and Patton, *The Patton Papers*, 781.
128. "Necessity for Horsed Cavalry under Modern Conditions," 251.
129. Crozier, *Ordnance and the World War*, 290.
130. Scott, "Progress in Cavalry Mechanization: Scout Car Developments," 281.
131. Herbert R. Rifkind, *The Jeep: Its Development & Procurement under the Quartermaster Corps, 1940–1943* (London: ISO Publications, 1943), 15.
132. Karl K. Probst with Charles O. Probst, "One Summer in Butler—Bantam Builds the Jeep," *Automobile Quarterly* 14, no. 4 (1976): 431–438.
133. Ibid., 433.
134. The average weight of a Percheron is 1,500 pounds. (See Galtrey, *The Horse and the War*, 123.) Similarly, draft horses in 1908 had also been calculated to be 1,500 pounds. (See Charles Hayward, "How the Horse and It's [sic] Load Wear Out Roads," *The Automobile*, June 18, 1908, 843.) This weight has not changed for the average Percheron today.
135. Bill Munro, *Jeep: From Bantam to Wrangler* (Marlborough, Wiltshire, U.K.: Crowood Press Ltd., 2000), 14–15.

136. Information about the Percheron was obtained in an interview on August
     31, 2005 with the following Cornell University veterinarians who special-
     ize in horses: Judy L. Urban (thirty years' experience in horse surgery),
     Ann Townsend-Poors (twenty years' experience in horse anesthesia), and
     Margie Vail (ten years' experience in horse surgery).
137. Munro, *Jeep: From Bantam to Wrangler*, 13.
138. *Automobiles: Jeep*, A & E Television Networks, 1996.
139. Rifkind, *The Jeep*, 15.
140. Scott, *Essential Military Jeep*, 11.
141. Ibid., 12. Also in Schreier, "Born for Battle," 32.
142. Zita Elaine Lambert and Robert John Wyatt, *Lord Austin: The Man* (Lon-
     don: Sidgwick & Jackson, 1968), 123.
143. John Underwood, *Whatever Became of the Baby Austin?* (El Monte, CA:
     R & L Press, 1965), 5.
144. Ibid. See also Lambert and Wyatt, *Lord Austin: The Man*, 126. Lambert
     and Wyatt calculate the length as 105 inches.
145. Underwood, *Whatever Became of the Baby Austin?*, 5.
146. Lambert and Wyatt, *Lord Austin: The Man*, 126.
147. Underwood, *Whatever Became of the Baby Austin?*, 418.
148. Probst with Probst, "One Summer in Butler," 431.
149. Ibid., 433.
150. Ibid., 435–438.
151. Ibid., 436.
152. Lieutenant George M. White, "Cavalry's Iron Pony," *The Cavalry Journal*
     50, no. 2 (March–April 1941): 85.
153. Ibid., 88.
154. Lieutenant Colonel John A. Considine, "Sixth Cavalry-(Horse Mecha-
     nized) Fort Oglethorpe, Ga." *The Cavalry Journal* 50, (January–February
     1941): 87.
155. White, "Cavalry's Iron Pony," 86.
156. Ibid.
157. Captain Bruce Palmer, "The Bantam in the Scout Car Platoon," *The Cav-
     alry Journal* 50, no. 2 (March–April 1941): 89.
158. Staff Correspondent, "The Sixth Cavalry in the Fourth Corps Maneuvers,"
     The Cavalry Journal 49, no. 219 (May–June 1940): 198.
159. Palmer, "The Bantam in the Scout Car Platoon," 89.
160. Phillips, "The First Motorized Cavalry," 10.
161. Palmer, "The Bantam in the Scout Car Platoon," 90.

162. Ibid.

163. Ibid.

164. Ibid., 91.

165. Major John Hughes Stodter, "Radio Equipment for Horse Cavalry," *The Cavalry Journal* 50, no. 3 (May–June 1941): 68–69.

166. Palmer, "The Bantam in the Scout Car Platoon," 89–91.

167. Major Albert Whipple Morse, Jr., "Stables, New Type," *The Cavalry Journal* 50, no. 3 (May–June 1941): 75.

168. Ibid.

169. "The Cavalry School of 1943," *The Cavalry Journal* 52, no. 1 (January–February, 1943): 85–86.

170. John L. S. Daley, "From Theory to Practice: Tanks, Doctrine, and the U.S. Army, 1916–1940" (PhD diss., Kent State University, 1993), 440.

171. Ibid.

172. Colonel Edgar S. Stayer, "The Year's Advancement in Military Motor Transport," *The Quartermaster Review* 12 (Jul–Aug 1932): 33–37.

173. "Fundamentals of Cavalry Training Policy," 184.

174. E. Risch, *Quartermaster Support of the Army: A History of the Corps, 1775–1939*, CMH Pub, Vol. 70–35 (Washington, DC: Center of Military History, U.S. Army: For sale by the Supt. of Docs., U.S. GPO, 1989), 718.

175. The Federal Trade Commission ruled that the name "jeep" was first used on a vehicle by an unidentified soldier or noncommissioned officer after the newspaper comic strip character (*Federal Trade Commission Decisions* 44, 572, cited in Frederic L. Coldwell, *Selling the All-American Wonder: The World War II Consumer Advertising of Willys-Overland Motors, Inc.* [Lakeville, MN: USM Inc., 1996], 98). The term "jeep" came from a Popeye comic strip published in the *Daily Times* on April 22, 1940. The versatile dog Eugene uttered "jeep, jeep, jeep," and a sergeant named James T. O'Brien liked the term and painted it on his vehicle. Several weeks later the name became common in the military (Ray Cowdery and Merrill Madsen, *All-American Wonder: Information Regarding the History, Production, Features and the Restoration of Military Jeeps, 1941–1945* [Rogers, MN: Victory Publishing Limited, 1993], 43). There is, however, a predominant misconception that the term "jeep" came from the abbreviation "GP," which stood for "General Purpose." However, it appears that GP most likely came from Ford's Parts Numbering System. The prefix G stood for "Government" and the prefix P stood for "80-inch wheelbase Reconnaissance Car." Typically, Ford would use the first digit of the prefix to indicate the year the part was made, i.e., prefixes were "8," for 1938, or "9," for 1939, etc. In the

case of the jeep, Ford used a generic letter G, most likely to indicate that it was a government model since its specifications were standardized and did not change as often as those for commercial cars, whose models varied on a yearly basis. The second prefix, which represents engine power, i.e., "1" for 85 H.P. V-8 or "2" for 60, etc., uses P to refer to the 80-inch wheelbase (Cowdery and Madsen, *All-American Wonder*, 21). Thus while the name "jeep" did not come from GP, and most likely came from a comic character, the Federal Trade Commission ruled that "it is impossible to state with certainty just when the name 'Jeep' was first applied" (*Federal Trade Commission Decisions* 44, 572–90, cited in Coldwell, *Selling the All-American Wonder*, 101). Although it was decided with a fair amount of certainty that when the *Washington Post* published a story on March 16, 1941 about this cross-country machine and referred to it as a "jeep," the name stuck (Probst with Probst, "One Summer in Butler," 436).

176. W. E. Butterworth, *Soldiers on Horseback: The Story of the United States Cavalry*, 1st ed. (New York: W.W. Norton & Company, Inc., 1967), 127.

177. Cowdery and Madsen, *All-American Wonder*, 46.

178. Erwin Rommel and Basil Henry Liddell Hart, ed., *The Rommel Papers*, with the assistance of Lucie-Maria Rommel, Manfred Rommel, and General Fritz Bayerlein, translated by Paul Findlay, 1st American ed. (New York: Harcourt, Brace and Company, 1953), 292. Kufra Oasis is some 500 miles south of Tobruk, deep in the Sahara Desert. For further details on Stirling's raiding areas, see Arthur Swinson, *The Raiders: Desert Strike Force* (New York: Ballantine Books Inc., 1968), 88–89.

179. Cowles, *The Phantom Major*, 71.

180. The traditional British commando unit, the Brigade of Guards, was large, typically consisting of about 200 men. This ponderous unit would approach German targets from the Mediterranean Sea, a strategy that gave their presence away even before they reached the North African coastline.

181. *Cavalry Service Regulations, 1914*, 260.

182. United States, War Department, Office of the Chief of Staff, *Cavalry Drill Regulations, 1916*, War Department Document No. 340 [hereafter cited as *Cavalry Drill Regulations, 1916*] (Washington, DC: GPO, 1917), 196–199.

183. Ibid., 196–98; 224–225.

184. Swinson, *The Raiders: Desert Strike Force*, 116.

185. *Cavalry Service Regulations, 1914*, 261.

186. Lee Kennett, *GI: The American Soldier in World War II*, 1st ed. (New York: Charles Scribner's Sons, 1987), 107.

187. Dwight D. Eisenhower, *Crusade in Europe* (New York: Doubleday and Company, Inc., 1948), 163–164.
188. Lyman M. Nash, "The True History of the Ugly," in *Military Jeeps 1941–1945* (Hong Kong: Brooklands Book Distribution Ltd.), 48. Reprinted from *The American Legion Magazine* (February 1967).
189. D. Colt Denfeld and M. Fry, *The Indestructible Jeep*, Ballantine's Illustrated History of the Violent Century, Weapons Book, Vol. 36, 126. Also in *The Unstoppable Soldier*, produced by John V. Coscia and William Stephens, written by William Stephens, Global Television Network, ltd., 1996.
190. See examples shown in the films *The Unstoppable Soldier* and *Automobiles: Jeep*. Also, see Cowdery and Madsen, *All-American Wonder*, 46–51; Denfeld and Fry, *The Indestructible Jeep*, 6–16, 72–73, 86–87, 112–145; Nash, "The True History of the Ugly," 48; A. Wade Well, *Hail to the Jeep* (New York: Harper & Brothers, 1946), 1–7; R. M. Clarke, *Jeep Collection No. 1* (Hong Kong: Brooklands Book Distribution Ltd., n.d.).
191. See the films *The Unstoppable Soldier* and *Automobiles: Jeep*.
192. Well, *Hail to the Jeep*, 43–46; Denfeld and Fry, *The Indestructible Jeep*, 73.
193. Ernie Pyle, *Brave Men* (New York: Grosset & Dunlap, 1945), 211.
194. Carole Landis, *Four Jills in a Jeep* (New York: Random House, 1942), vii–5.
195. George C. Marshall, Larry I. Bland, and Sharon Ritenour Stevens, *The Papers of George Catlett Marshall* (Baltimore, MD: Johns Hopkins University Press, 1981), 575.
196. Mildred Hanson Gillie, *Forging the Thunderbolt: A History of the Development of the Armored Force* (Harrisburg, PA: Military Service Publishing Company, 1947), 219.
197. Well, *Hail to the Jeep*, 2–3; Denfeld and Fry, *The Indestructible Jeep*, 129.
198. Denfeld and Fry, *The Indestructible Jeep*, 129.
199. Well, *Hail to the Jeep*, 4.
200. Jim Lucas, "Occupation of the Russells," in *The United States Marine Corps in World War II*, ed. S. E. Smith (New York: Random House, 1969), 364.
201. The Field Ration "C" consisted of three 12-ounce cans of ready-to-eat meals, designated M-Units, and three cans, each containing bread, coffee, sugar, and chocolate, known as B-Units. One meat meal, or M-Unit, was roast beef hash; one was beef stew; and the third was a meat (part-pork) and bean mixture. See "Editor's Mail," *The Cavalry Journal* 50, no. 3 (May–June 1941): 46.
202. Ernie Pyle and David Nichols, ed., *Ernie's War: The Best of Ernie Pyle's World War II Dispatches*, 1st ed. (New York: Random House, 1986), 111.

203. James Tobin, *Ernie Pyle's War: America's Eyewitness to World War II* (New York: Free Press, 1997), 94.
204. Bill Mauldin, *Up Front* (Cleveland: World Publishing Company, 1946), 212.
205. Bill Mauldin, *Back Home* (New York: William Sloane Associates, 1947), 72.
206. Pyle, *Brave Men*, 259.
207. Ernie Pyle, *Last Chapter* (New York: Henry Holt and Company, 1945), 137.
208. Ibid., 51.
209. "One of the Faithful," "Faith in and a Doctrine for the Cavalry Service," 227.
210. Pyle, *Brave Men*, 212.
211. Denfeld and Fry, *The Indestructible Jeep*, 129.
212. Nash, "The True History of the Ugly," 53; Denfeld and Fry, *The Indestructible Jeep*, 141.
213. Cowdery and Madsen, *All-American Wonder*, 46–47.
214. Stewart, "Mechanized Cavalry Has Come to Stay," 489.
215. *Cavalry Service Regulations, 1914*, 9.

Chapter 4

Case Three:
Domesticating the Jeep:
The Philippines

Despite the dramatic decrease in the use of the horse, the jeep immortalized the cavalry spirit through mechanical means. Even today, any highly mobile army unit that uses transport, such as light armor or helicopters, is designated "cavalry." The cavalry's strategy of stealth, speed, mobility, and dramatic offensive attacks shaped the form and functionality of the World War II jeep.

This chapter examines a similar shaping of the jeep conducted in an entirely different setting, the Philippines after World War II. The jeep began as a peripheral object in the Philippines in two ways: first, it was a military object from a different country, and second, it was a mechanical form of transport. Similar to the horseless-carriage days at the turn of the century in the United States, the Philippines at the advent of World War II relied largely on muscle power. How did the jeep, an alien product in the eyes of Filipinos, attain universal appeal in an environment so foreign to its origins?

The domestication of the jeep in the Philippines constitutes the third case study in understanding the role of the horse concept in facilitating motorization. Many historical records were destroyed during the war; thus accounts of everyday Filipino life tend to be documented by foreigners.[1] While their perspective might be biased, many of them arrived having witnessed wars in other countries; thus they provide an informed view of the circumstances in which the Philippines emerged from the war. Historical accounts of some expatriates express a fondness for the Philippines, and their reminiscences may sometimes sound like those of local residents.

Western observers would simultaneously highlight the peculiarities and familiarities of the Philippines relative to the modern world; after all, the jeep is a product of the West. Photographs, paintings, and observations from these travelers capture nuances that a local person might take for granted. The extensive, four-volume *Philippine Commission Report* of 1900–1901, with transcripts of interviews conducted by the American delegates with key local leaders, proved to be a useful primary source in gleaning a sense of the state of roads, infrastructure, culture, and practices prior to the arrival of American influence.

Since the localized jeep, or jeepney, continues to be used today, I have also examined contemporary accounts and reflections of a local literary scholar, Emmanuel Torres, known for articulating the sentiments of the general public, including marginal groups in Philippine society. He looked specifically at the origins of the jeepney and discussed the mind-set of jeepney drivers during the vehicle's heyday in the 1970s. I have supplemented this account with phone interviews conducted in the Ilocano and Tagalog (Filipino) dialects with fifteen jeepney drivers working in the Philippines at the time of this writing. They were selected because they had worked in Manila and various rural areas in Luzon, the northern island group of the Philippines. Some have driven horse-drawn carriages. These drivers provide insights into current thinking about the jeepney and its slow descent into obsolescence.

While conducting interviews on the phone may seem less effective than conducting face-to-face interviews, in this case, phone interviews were advantageous. When jeepney drivers were approached with a tape recorder or a survey questionnaire, they tended to shy away.[2] Many of them were uncomfortable with having to read and write. Tape recorders were particularly unnerving because of the long history of suppression of freedom of speech during the thirty-year Marcos regime. The phone interview setting allowed drivers to be more at ease, given the prevalence of cellular phones in the Philippines, and also prevented the problem of onlookers flocking around an interview setting.

Finally, I have traced the life stories of the pioneers and original manufacturers of the jeepney using accounts and interviews conducted by various Philippine magazine writers. The original manufacturers of the jeepney are deceased, and in the absence of books and other primary source materials on the topic, magazines provide an alternative resource. These magazines target middle-class readers and, hence, provide a reasonable approximation of the types of issues and concerns that dominate mainstream thinking. Some of the articles were published transcripts of interviews.

This chapter shows the jeep's endurance as a mechanical surrogate for the horse. While the previous chapters discussed *why* the jeep assumed its particular form and functionality, this chapter focuses on *how* the jeep attained its universal appeal in the Philippines. The jeep evolved into a custom-built, elaborately ornamented passenger vehicle that continues to provide cheap and convenient transport for Filipinos. The equine legacy of the Philippines in many ways paved the way for the ease with which a localized jeep became part of the Philippine landscape.

SURPLUS JEEPS

After the war, the jeep's evolution took various interesting turns. American manufacturers petitioned the U.S. government to protect their markets from being flooded with surplus wartime goods. Willys-Overland, the sole manufacturer of jeeps by 1945, argued that returning the jeeps to the United States would hurt the company's postwar business, making it difficult to generate jobs for returning GIs.[3] Additionally, Willys testified before the

U.S. House of Representatives that jeeps were designed for battle and, thus, for safety reasons, they would need to be modified for civilian use.

Consequently, most wartime jeeps were kept from being brought back for sale in the United States. What happened to the surplus jeeps left in Allied countries, however, was another matter. In the interest of developing new markets for spare parts and other downstream supply businesses, jeeps were left behind. The supposed concern for civilian safety in using a military-designed vehicle became a superfluous issue outside the United States.

The stripped-down jeep in the Philippines was easily localized into an elaborately ornamented passenger vehicle used largely by the public. The jeep in the Philippines became the "jeepney"—a term perhaps derived from "jitney," the five-cent-fare auto that came into use in Los Angeles in 1914,[4] although this association was not mentioned in any widely cited literature on the jeepney. The jeepney replaced the *calesa*, the horse-drawn carriage used for public transport before the war. While the lack of transport in postwar Philippines might have motivated the immediate adoption of the localized jeep, its ornate look and feel could not be explained by necessity alone.

THE HORSE LEGACY

A working Philippine cavalry appears to have existed as early as 1898. The Philippine cavalry put up strong resistance to the Americans as part of a four-year insurrection for independence[5] following the American seizure of the Philippines during the Spanish-American War. No U.S. cavalry was assigned in the Philippines in 1898, but on March 2, 1899, Congress ratified three cavalry units, two of which were organized in the Philippines: one squadron composed of Americans then in the Philippines and the other composed entirely of Filipinos.[6]

The Philippine cavalry unit under American supervision was perhaps organized as a countermeasure against resistance forces. On April 23, 1899, the American cavalry under Major James Franklin Bell suffered heavy casualties, including the death of Colonel John M. Stotsenberg, at the hands of Filipino hero General Gregorio del Pilar.[7] The preliminary report of the Philippine Commission to the President of the United States on November 2, 1899, expressed grave concern about the Filipino resistance:

> The insurgents were insolent to our guards and made persistent and continuous efforts to push them back and advance the insurgent lines farther into the city of Manila. It was a long and trying period of insult and abuse heaped upon our soldiers, with constant submission as the only means of avoiding an open rupture. [...] Rumors were always prevalent that our army will be attacked at once.[8]

When the former chief of the northern province of the Philippines, Senor Angel Lopez, testified before the committee earlier that year on May 8, his repeated assurances of holding no insurgents[9] in his area indicate the issue to be of vital concern to the colonizers.[10] The committee was composed of President Schurman (in the chair), Admiral Dewey, Colonel Denby, and Professor Worcester, commissioners, and Mr. John R. MacArthur, secretary. The following is an excerpt of the inquiry between Schurman and Lopez:

> Q. How do the people of Union and Ilocos [one of the northern provinces in Luzon] stand at the present time; how are they affected toward the Americans and toward the insurrectos?
>
> A. In Vigan [capital city of Ilocos Sur] there is no insurrection; there are Philippine troops, but they are not insurgent troops.
>
> Q. Where do those troops come from?
>
> A. They are from Vigan itself; they are recruits, new troops, reserves.[11]

However, it was not just the troops that interested the committee. Worcester asked Lopez:

> Q. Is gold found in the sand or in the rocks?
>
> A. In the forests and in the mountains, in the rocks.
>
> Q. Are there wild people in the mountains?
>
> A. No.

Q. Can one travel on horseback through this province during those rains?

A. Well, even in time of rain you can travel on horseback or in a calesa or a carriage. They have got good roads.[12]

Lopez described a well-developed road infrastructure in 1899. Even when it rained for one week continuously, the roads were passable[13] (Figure 57). Horses were an integral part of Filipino daily life, such that efforts were made to create good roads purposely designed to withstand heavy monsoon rains. Horses were used to haul timber for building schools and houses[14] and to haul sheaves of rice.[15] The major means of transport also used what looks like a local version of a horse buggy (see Figures 58, 59, and 60).

FIGURE 57. Flooded Manila street.

Source. Report of the Philippine Commission to the President, 1900, vol. 3 (Washington, DC: GPO, 1901), Plate 16.
Note. Caption reads "A street of Manila in rainy season."

FIGURE **58.** Horse-drawn carriages in the Philippines.

Sources. (A) *Report of the Philippine Commission to the President, 1900,* vol. 3 (Washington, DC: GPO, 1901), Plate 45. (B) *Report of the Philippine Commission to the President, 1900,* vol. 3 (Washington, DC: GPO, 1901), Plate 46.
Note. Captions read (A) "Filipino equipage, common in Manila." (B) "A carromata."

FIGURE 59. Pony-drawn carts.

Sources. (A) *Report of the Philippine Commission to the President, 1900,* vol. 3 (Washington, DC: GPO, 1901), Plate 44. (B) *Report of the Philippine Commission to the President, 1900,* vol. 3 (Washington, DC: GPO, 1901), Plate 47.
Note. Captions read (A) "Filipino pony and cart." (B) "Cart commonly used by the natives."

Figure **60.** Philippine horses at work.

Sources. (A) *Report of the Philippine Commission to the President, 1900,* vol. 3 (Washington, DC: GPO, 1901), Plate 6. (B) "Public Works and Edifices," Paper No. 11, *Report of the Philippine Commission to the President, 1900,* vol. 4 (Washington, DC: GPO, 1901), Plate 3.
Note. Captions read (A) "Hauling sheaves of rice." (B) "Hauling a huge timber for the school of Mercedes, in course of construction (Zamboanga)."

The yearly horse shows in Manila prior to World War II were popular and well respected, even among American cavalry officers. The Philippine horse tradition permeated Filipino life such that Philippine folklore and legends featured a horse creature that haunts travelers. *Tikbalang*, a reverse-centaur with a horse head and a human body, dominates many Philippine folklore stories. He is known to sit and smoke a pipe on top of large trees, virtually unseen except for the smoke rising out of the trees. The *tikbalang* preys on travelers, especially at night, by giving false directions and getting them lost in the forest. Adults have been known to use *tikbalang* stories to frighten children from straying far from home and to ensure their return home by nightfall. Whereas horses have been known to find their way home, the *tikbalang* is the reverse of what the horse symbolizes in everyday life, a creature that disorients people to keep them from finding their way home.

When World War II began, the Philippine transportation system still depended largely on horses. It is interesting, however, that although the jeep became a ubiquitous transport in the form of the jeepney, it was never adopted as a farm implement, just as horses were never adopted in the Philippines for farm use. Despite a largely agricultural economy, farm applications of the jeep never materialized. Filipinos used water buffaloes, not horses, to till the land. The preexisting practice of using horses largely for transport continued during the years prior to the arrival of the jeep. Thus one could argue that preexisting practices associated with the use of horses for farm applications did not exist; hence the jeep, being a mechanical equivalent of the horse, was also not used in agriculture. At the same time, it is also likely that the high water level in rice paddies rendered the water buffalo a more effective farm animal than the horse.

WORLD WAR II IN MANILA

When James Bertram Reuter came to the Philippines as a Jesuit scholar in 1938, he observed, "Manila was quiet. Caratellas: more caratellas than cars."[16] "Caratellas," properly spelled in Spanish as "caretellas," and "karitela" (singular) or "mga karitela" (plural) in the vernacular language, is the local term for horse-drawn carriages. However, a more popular term currently used would be "*calesa*," although "carromata" (Spanish version) or "karomata" (local term) has been used, particularly in the Ilocano dialect in the northern provinces of the Philippines. Transportation in Manila, even in the late 1930s, had not changed much from 1901 (Figure 61); however, the peace and quiet Reuter noted in Manila along with its bucolic carriages soon changed as World War II ensued.

While the Philippines appeared to have had fairly robust roads by the turn of the century, the Japanese occupation destroyed much of the transportation infrastructure. The Japanese halted virtually all economic activities in the Philippines. The number of work animals decreased significantly. The Japanese confiscated all horses, cars, trucks, and other means of transportation.[17] Toward the end of the war, highways and roads were virtually unusable. Edward Woolbright, an American entrepreneur,

FIGURE 61. Manila in 1900.

Source. Report of the Philippine Commission to the President, 1900, vol. 3 (Washington, DC: GPO, 1901), Plate 10.
Note. Caption reads "The Bridge of Spain (Manila)."

described the state of roads and highways in one of the cities in the southern islands of the Philippines in late October 1944:

> See the Army was there…hundreds and hundreds of trucks and all kinds of vehicles…tanks…bulldozers. All of them up and down the roads…The roads were just muck. You couldn't even walk down the roads…[we had to go] in a jeep. Mudholes two or three feet deep in the town and all.[18]

When Manila, the capital of the Philippines, emerged from the war, mudholes were not the only problem. After massive bombing by American forces, burned bodies and large pieces of debris from buildings were strewn everywhere. Herbert Zipper, a Viennese composer and conductor who had arrived in the Philippines in 1939, described the remains of the war:

> We were in Manila when liberation came, when bombs and shells tore the city apart but again set it free. We saw Manila in flames. We walked through the miles of scorched earth, through districts where a few days before homes, schools, churches and hospitals stood. We saw the ugly grimaces of war: unsupported walls, marble stairs leading nowhere, bathtubs and toilets as if suspended in mid-air, trees and shrubs charred by fire. And we smelled the stench of war. No one knows how many thousands of Manila's people perished in February 1945, their putrefied bodies unburied, strewn over dozens of square miles.[19]

Zipper went on to describe Manila as lacking in any "amenities of civilized life" after seeing the luxury and sophistication of a European-looking city destroyed.[20] There was no water, food, electricity, gas, telephone, or transportation system. An American historian described how Japanese soldiers hid in Manila and the consequent battle with American soldiers nearly reduced the city to ashes:

> Perhaps the foe had thought to set one in Manila, for fires like those that ruined old Moscow in Napoleon's day broke out everywhere, and in the older sections Japanese soldiers fought to the last from house to house and block to block, while on Corregidor our paratroopers had to burn and slaughter the adversary out of

his hide-outs one by one. It was a bloody, ghastly business and took a long time, the greater part of February, to clean Manila.[21]

Paul D. Perrine, an American GI who stayed in the Philippines after the war, confirmed this account, and described the devastation in Manila not only in terms of destroyed buildings but also in terms of the desolation permeating the city:

> The Japanese burned several buildings and killed an estimated 100,000 civilians. Many were also killed by the U.S. military and bombing, as the fighting took place in heavily populated residential neighborhoods. So it was a sad, deplorable sight that greeted us, as we were able to look around after it was liberated. The city was actually liberated; we were able to get in and look around. I believe General Eisenhower visited here shortly after the end of World War II, and he said that of all the cities that he had seen, he thought that this was the second most destroyed city, next to Warsaw.[22]

A photograph published in *The Cavalry Journal* in 1945 hinted at the devastation in Manila (Figure 62).

When Lyle A. K. Little arrived in Manila in early 1945 as part of General Douglas MacArthur's staff, he also made the comparison to Warsaw. He specifically described the devastation of the transportation system but noted the ingenuity of the people who scraped to make what he describes as "marvelous things with nothing."

> When I arrived in Manila it was in absolute shambles, second only to Warsaw as the most devastated city in World War II. All the government buildings were in ruins. There was no electric- ity…The streetcars were wreckage and piled one on top of the other. It was a very, very depressing sight…There were a few cars on the streets. Paved streets were nonexistent. They had turned to gravel and mud and dust. But there were a few civilian cars operating, some of them with the clumsy old charcoal burners mounted on the outside rear of the cars, which made them look very grotesque. They were burning charcoal seeking to convert it to gas—not a recommended way. They were quite ingenious these people. For all their gauntness and privation and their lack

FIGURE 62. Manila in 1945.

Source. "Cavalrymen Re-enter Manila," *The Cavalry Journal* 54, no. 2 (March–April 1945), 44. Reprinted with permission of the U.S. Cavalry Association.

of everything, they were ingenious at improvising and did marvelous things with nothing.[23]

This genius for improvising came into full flower with the local adaptation of surplus jeeps left in various military depots. Citizens in many Allied countries, confronted with a glut of surplus jeeps, had to figure out what to do with these "unsafe" vehicles barred from going home.

THE AUTO CALESA: THE PERSISTENCE OF PRACTICES

The devastation of roads and highways, together with the confiscation of virtually all means of transport by the Japanese during their occupation of the Philippines, made the durable all-terrain vehicle a ready solution to the bomb-pocked roads. The rugged U.S. Army jeep instantly became a status symbol after the car-less years of the Japanese occupation.

Civilians invited to ride in one of these army jeeps felt, according to a foreign expatriate, "elegant and superior" in the midst of the rubble.[24] Thus the use of converted jeeps as passenger vehicles at this time had practical as well as symbolic appeal, although their panache emerged more from the dire conditions of a war-torn country in desperate need of virtually everything.

On March 10, 1945,[25] almost immediately after the liberation of Manila, the wartime jeep was officially declared a "passenger automobile" by the Philippine Motor Vehicle law instituted under Executive Order No. 34, enacted by the Senate and House of Representatives of the Philippine Congress.[26] At this time, a motor vehicle registered to carry more than ten passengers was considered a "passenger truck" rather than a "passenger automobile."[27] Thus the jeep, a "passenger truck," out of sheer necessity emerged as an immediate solution to the problem of transportation. How the jeep was used and how it evolved, however, were far more complex.

Trudl Dubsky Zipper, an expatriate artist, described the existence of a jeep-like "baby bus" with an Austin Seven chassis that was presumably the same Austin from which the jeep originated.[28] The "baby bus," also referred to as an auto calesa, or AC (literally "automated horse-drawn carriage" but the term also plays on the AC as air conditioner because of its open body windows), closely resembled the jeep. She painted a watercolor depicting an auto calesa, shown in Figure 64.

Some auto calesas, according to the artist, were hidden well enough to escape the eyes of the Japanese. After the war, they emerged from hiding and were put to immediate use. Examining the watercolor, one can see that the number of passengers the auto calesa was carrying far exceeded the capacity of the Austin Seven, a practice that continues today. These early auto calesas could have arrived in the Philippines at the beginning of the war, that is, at the beginning of the 1940s rather than in the 1930s.[29] The 26th U.S. Cavalry, also called the Philippine Scouts, together with tanks and scout cars, was already training in the Philippines at this time, several months before the bombing of Pearl Harbor. The cavalry trained "with the idea in mind that it was actually at war."[30]

Indeed, as early as July 26, 1941, George Marshall, in a secret letter to General Douglas MacArthur, established the Philippines as the Headquarters of the United States Armed Forces in the Far East,[31] later known as the USAFFE. It was the 26th Cavalry Division that first fought the Japanese invasion force that landed on the coast of the Philippines on December 10, 1941.[32] The auto calesa referred to by the watercolor artist most likely came from one of these early regiments. Nevertheless, the local reference was consistent—a horse-drawn carriage that happens to be automated. The horse concept remained central whether Trudl's "baby bus" was indeed an Austin car or a Bantam jeep.

Several references have connected the auto calesa with the jeep. The auto calesa was described as seating eight to ten people (although the artist admitted that the eight-to-ten capacity meant, in practice, a *minimum* of eight to ten, as is evident in the illustration), the exact benchmark established by the Philippine Motor Law of 1945 that specifically mentioned the jeep as a passenger vehicle. Indeed, a Filipino art critic and poet used the term "auto calesa" to refer specifically to converted surplus jeeps from which the jeepney originated:

> In a sense the jeepney is nothing new. It is really a motorized version of that once King of the Road, the horse-drawn calesa. In fact, as if to establish the connection, the early handcrafted short-body jeepney which seats no more than 10 (three or four on each side of the main cab section, plus two up front beside the driver) is called Auto Calesa, or AC.[33]

Even later accounts of the auto calesa recognized it as none other than a basic wartime jeep, with no significant modifications to its passenger-carrying capacity. A writer described the early models of the jeepney as equivalent to the auto calesa:

> Because its first models were nothing more than motorized versions of the horse-drawn *calesa*, the small jeepneys which sat three or four people on each side were called AC or Auto Calesa. (This is probably the reason why most jeepneys carry miniaturized

chrome equestrian figures on their hoods—to remind us of the hardy animals they have displaced.)[34]

The auto calesa came from an equine tradition in which its use, not just its look, evoked the heritage of horse-drawn carriages. The term "auto calesa" means automated horse-drawn carriage, a concept in which the horse remains integral, in contrast to the term "horseless carriage," where the horse has been literally replaced by a motor. In the Philippine concept of a motorized vehicle, the horse element remained central, despite the absence of the physical horse.

Even in practice, people used the auto calesa like a horse-drawn *calesa*. The style of communal use found in the horse-drawn *calesa*, as represented in the 1945 watercolor, was replicated in the auto calesa. Observe the similarity in riding practices of the *calesa* versus auto calesa passengers (Figure 63 versus Figure 64). Note that these watercolors were drawn from similar side views by an expatriate who captured the distinctive practice of shared riding.

As the watercolors illustrate, the way passengers rode in the auto calesa closely parallels the manner in which the original *calesa* was used. The sardine-packed communal seating of the *calesa* was simply transferred to the auto calesa. The method of payment also followed the honor system of the horse-drawn *calesa*, by which passengers ride first and pay after.[35] Jeepneys later continued this tradition.

When the rear of the auto calesa was extended to accommodate a greater number of passengers, the commercialization of what would become known as the jeepney began. The transformation of the auto calesa into the jeepney was described as follows:

> [The auto calesa] differs from its later development, the assembly-plant type with body extended to accommodate more passengers, from 14 to 18, including the two up front, known as the Public Utility Jeepney, or PUJ.[36]

According to this writer, passenger capacity seems to determine the difference between an auto calesa and a jeepney. Nevertheless, the practice

FIGURE **63.** *Calesa.*

Source. "The Carretela," by Trudl Dubsky Zipper. From Trudl Dubsky Zipper and Herbert Zipper, *Manila 1944–45: As Trudl Saw It* (Santa Monica: Crossroads School, 1994), 77. Reprinted with permission from the Crossroads School.

of cramming in as many people and as much luggage as possible remained in force regardless of the means of transport.

The sheer lack of basic necessities in the aftermath of war made surplus wartime jeeps an attractive, immediate solution to the problem of transportation. The motivation of American businesses to develop foreign trade also encouraged the use of U.S.-made machinery and goods. Eddie Woolbright, an American entrepreneur who arrived in the Philippines in 1944, described his opening of a spare parts store in the southern part of the Philippines: "We could do anything the Filipinos could do: we could buy, sell, put up a business. So it was a great life."[37] Parity rights were instituted in 1946 by the United States as a condition of Philippine independence: American citizens had the same rights as Filipino citizens; that is, they could live in the Philippines, do business,

FIGURE **64.** Auto calesa.

Source. "Auto Calesa," by Trudl Dubsky Zipper. From Trudl Dubsky Zipper and Herbert Zipper, *Manila 1944–45: As Trudl Saw It* (Santa Monica: Crossroads School, 1994), 59. Reprinted with permission from the Crossroads School.

and exploit the country's natural resources.[38] It was initially difficult for Filipinos to compete with Americans in the war-torn economy.

The wartime jeep, with its ease of maintenance and ready availability of spare parts, increasingly offered a permanent solution to Philippine transportation problems. Furthermore, local adaptation made the jeep a part of the Philippine landscape. While the use of converted jeeps as a form of mechanized horse-drawn carriage may appear to have simply occurred, stories abound about who actually conceived the idea. As was the case with the story of the jeep in the United States, the attribution of the jeepney's origins to one individual will perhaps continue to be a matter of contention.

JEEPNEY KINGS: THE LEGACY OF THE HORSE

Filipino scholars agree that the jeepney originated from the surplus wartime jeeps left behind by the U.S. military.[39] However, similar to the origin of the jeep, many claims have been made about who first devised the concept. A group of Filipino inventors recognized Clodualdo (Clod) Delfino, a bandleader and composer of popular music, as the first person to come up with the idea.[40] He was out of work after the war and decided to convert one of the leftover wartime jeeps into a passenger vehicle. The jeep at that time had a canvas top. Delfino provided the initial sketches of the "first" jeepney (Figure 65), which for all practical purposes resembles Trudl's auto calesa (Figure 64).

FIGURE **65.** Sketch of the "first" jeepney.

Source. Emmanuel Torres, *Jeepney* (Quezon City, Philippines: GCF Books, 1979), 39. Reprinted with the permission of GCF Books.

Delfino displayed the name "jeepney" on his windshield. Just as the name "jeep" became commonplace in the military because a sergeant supposedly painted the name on his vehicle,[41] the "jeepney" name also stuck throughout the following decades, reportedly as a result of Delfino's initiative. Delfino quit after six months when the entertainment industry began to revive. He reestablished his band and sold his jeepney in 1946. By this time, numerous jeepneys offered similar transportation services.[42]

Whether Delfino indeed invented the jeepney concept requires far more convincing evidence than the personal testimonies of a handful of people. Certainly the use of auto calesas appears to have been well underway immediately after the war—and possibly even earlier. The commercialization of the jeepney, however, is a matter of less contention. It has consistently been attributed to two pioneers, both former carriage-factory workers. Just as carriage craftsmen in the first decade of the century played a critical role in the U.S. auto industry, these former carriage-factory workers, soon to become manufacturers themselves, created sustained mass appeal by capitalizing on the flair and flamboyance of the *calesa*.

The local adaptation of surplus wartime jeeps in the Philippines occurred at the grassroots level. Anastacio Francisco was born to a poor farm family and went to Manila in search of a better life. He began as an apprentice painter and was hired to work at the Mata Carriage Factory, where he met Leonardo Sarao, the second jeepney pioneer.[43] Together at the carriage factory, Francisco and Sarao worked as a team: Sarao lathed the wooden and metal parts of the carriages, while Francisco painted them. This background in carriage building later influenced the many expressions of equine culture found in jeepneys, similar to the way in which early horseless carriages in the United States came to be transformed by coach builders.

The commercialization of the jeepney—namely, the building of assembly-type jeepneys as opposed to the conversion of surplus wartime jeeps—began around 1947 when Francisco left the carriage factory to establish his own paint shop. A year later, most of the jeepneys and trucks roaming the streets of Manila were his "customers."[44] By this time, the jeepney began to evolve as a viable, permanent solution to the transportation needs of the city. Significant modifications to the short body of the jeep began. The rear portion of the wartime "passenger truck" was elongated to accommodate more passengers. The canvas top was replaced by an aluminum or iron roof to protect passengers from the elements. Although no two jeepneys were alike, a rough image of the transformation is shown in Figure 66.

By 1951 Francisco began to expand his painting business, branching out into body building and repair. The popularity of the jeepney accelerated in the 1960s and 1970s. Francisco by this time began manufacturing chassis frames and other metal components.[45] By 1983 there were 427,925 jeepneys, compared with only 16,714 buses[46] and a practically nonexistent train system.[47] Francisco's painting business helped transform a mass-produced vehicle into something unique and extravagant.

Sarao, on the other hand, specialized in the technical and manufacturing aspects of the jeepney business. Sarao's business also began in 1947 when he opened his own assembly shop, the same year Francisco opened his paint shop. Sarao's background came from the horse-drawn carriage industry. Following in his father's footsteps, Sarao began work as a rig driver before becoming a lathe operator.[48] Similar to Francisco, Sarao did not obtain a formal education beyond grade school.[49] By 1958 he was mass-producing jeepneys,[50] having built his own assembly plant by that time.[51] In the 1960s, jeepneys carrying the Sarao logo outnumbered all others seven to one.[52]

Francisco and Sarao dominated the jeepney business throughout the next several decades. Their commercialization efforts represented a fusion of the machinelike wartime jeep and the festive image of the horse-drawn carriage. A writer summarized Sarao's and Francisco's styles and backgrounds:

> Sarao's *cochero* [driver of horse-drawn carriage] background, like Francisco's, is readily recognizable in his jeepney designs. The horse motif is forever present. On the hood can be found three, six, even up to 14 stallion emblems, while horse tails and reins are painted on the sides.[53]

FIGURE 66. Transformation of the jeepney.

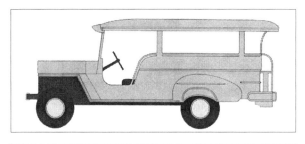

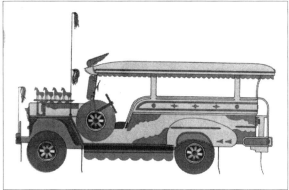

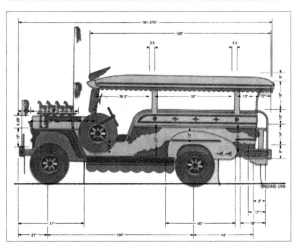

Source. Emmanuel Torres, *Jeepney* (Quezon City, Philippines: GCF Books, 1979), 42. Reprinted with the permission of GCF Books.

The tradition of decoration and flair in the *calesa* made its way into jeepney designs. Thus the use of the jeepney not only paralleled the use of the *calesa*; its look and style also mimicked the embellishments of Filipino horse-drawn carriages. The Philippine Institute of Hotel Administration described the manner in which the jeepneys were adorned like "decorated horse-drawn carriages prancing in the cobbled streets of old Manila."[54] The horse emblems found in the jeep signified not only the *cochero* background of the two pioneers but also, arguably, the pervasive role of the *calesa* itself in Philippine culture.

NEW KING OF THE ROAD

The horse emblem had become as indispensable an element of the jeepney as its mechanical parts. The iron sculpture of a standing horse on top of the hood, right above the grill guard, was the most popular chrome emblem after 1955.[55]

> The hood, which is about a meter square, is the special repository of all the creativity of the jeepney decorator. Almost mandatory here is the chrome horse standing upright on the hood, which may perhaps signify an awareness of the vehicle's "root," the horse-drawn calesa.[56]

It is interesting that the local writer describing the "mandatory" horse emblem of the jeepneys attributes its origin not to the U.S. military jeep but rather to the displaced local horse transport. What he observed was the family resemblance between the current jeepney and the *calesa*, particularly in the way its technological form mimicked that of the horse.

In the early days, the horses that pulled the *calesa* were embellished like kings, with little crowns on their heads.[57] Similarly, the hood of the jeep was embellished with chrome and décor that resembled the head of a *calesa* horse. This skeuomorphic practice, which George Basalla described as old elements persisting in new artifacts despite having no practical application,[58] played a critical role in transforming the foreign artifact into a familiar object. Indeed, these horse emblems and embellishments were so important that an entire industry emerged devoted solely to the jeepney's decorative regalia, paintings, upholstery, and all the additional elements that transformed a mass-produced frame into customized folk art on wheels.[59]

These embellishments made each jeepney one of a kind: no two jeepneys were ever identical—a far cry from their stringently standardized jeep predecessors. However, the horse emblem consistently appeared in each of these variations. The design and structure of the jeepney could even be modified according to the buyer's specifications.[60] Thus passenger capacity in jeeps ranged from 16 or 18 passengers[61] to as many as 22 to 30,[62] although this upper limit was more the exception than the rule. The horse emblem persists, however, throughout these customizations.

The singularity of each embellished jeepney provided the means to distinguish it from the ordinary and common vehicles of the day. While the decorations and embellishments supposedly attracted more passengers, they had more to do with the driver's desire for personal expression than with branding, given the lack of consistency across various vehicles. Dario Vega, a jeepney driver for nine years, spoke of the horse chrome decoration as his way of having "porma," or a sporty, dashing image.[63] Rupino Prestosa, a jeepney driver for fifteen years, likewise spoke of the horse chrome as "pang-arte," or artistic flair, artistic expression.[64] The majority of the jeepney drivers interviewed, about ten out of fifteen, mentioned one or both of these factors in describing the purpose of the horse chrome.

This need for personal expression also simultaneously corresponded to a shared universal value when the drivers spoke of seeing the horse chrome as a Philippine icon. Dario Vega described it as "simbolo ng Pilipinas," a "symbol of the Philippines." Still others described the horse as part of shared folk beliefs like the *tikbalang*, although in this case it is a good omen. Jojo Quines, who had been driving a jeepney for twenty years at the time of our interview, described the horse chrome as "swerte," or a "good luck charm," as well as "pamahiin," or an "auspicious symbol, superstitious belief."[65]

This paradoxical meaning of the particular and the universal is also expressed in the jeepney itself. No two jeepneys are alike, but commuters recognize jeepneys in general rather than a particular jeepney manned by a particular driver. Sarao and Francisco, however, branded many of the vehicles they produced, although one rarely sees their names these days because they have gone out of business. They are nevertheless still known as the "Jeepney Kings," whose brand remains immediately recognized by present-day jeepney drivers. Every jeepney driver interviewed knew the brand names Sarao and Francisco.

Still, many refer to the jeepney as the new "King of the Road," a title previously reserved for its predecessor, the *calesa*, which often displayed a big sign right above its roof proclaiming it "Hari sa Kalsada," or "King of the Road."[66] The jeepney drivers themselves came to embody this role when they spoke of their work as sovereign or self-governing. For instance, Alfredo Bodayong, 51 years old, had driven a nine-seater jeepney for twelve years when I interviewed him and planned to continue driving jeepneys despite the low pay, because, he reasoned in the Ilocano dialect,[67] "Awan gamin iti agbabaon," which translates as "because no one orders you around."[68]

The independence jeepney drivers relish has been further enhanced by their imagination. Abetted by various decorative representations that helped create a fantasy world, the jeepney driver added props to his vehicle that reminded him[69] of home and, at the same time, a place of adventure. The horse embellishment itself became the necessary prop for an imagined reality, a gateway from the grueling realities of long hours of work and the smog, dirt, and heat of city life to a heroic quest. A Filipino art critic and poet described the driver's sentiment:

> With saddle and stirrups (both simulating tooled leather), the gleaming "silver" or "white" horse standing in readiness makes a splendid symbol. It stands for what every driver wants his runabout to be, fleet and rugged. This is no puny Batangas [name of a local province in the northern part of the Philippines] horse but the heroic, muscular, tall variety seen in John Wayne movies,

or better yet, the Lone Ranger's Silver. This symbol makes it so much easier for the aficionado of Western movies to indulge in a favorite fantasy where the reality of traffic outside his windshield is the Wild, Wild West, the machine he drives a white steed, and he—Jun Aristorenas or Fernando Poe, Jr. [local action heroes], in a local version of a cowboy movie, or a *pancit palabok* [local noodle dish which connotes a mixture of things] Western.[70]

Indeed, Rupino Prestosa described the horse chrome, propped up typically with a spring, as a galloping horse, particularly evident when the jeepney starts moving. "Kasla agtartaray kin aglagtolagto nga kabalyo," or "It's like a running and jumping horse."[71]

Rather than having his jeepney take him somewhere far away, a jeepney driver imagined an escape fueled by his vehicle's various embellishments. In many ways, the jeepney drives like its military jeep predecessor, but the experience is similar to riding a horse. Jojo Quines, who had driven a *calesa* as a young lad and jeepneys for twenty years, observed that both the *calesa* and the jeepney lack shock absorption and ride roughly but that driving a *calesa* is easier because the horse has intelligence ("mautak," or literally, "smarts"). "You pull the reins and it runs! But one cannot do that with a jeepney."[72] Many of the jeepney drivers who had driven cars, trucks, and even *calesas* generally agreed that driving a jeepney was the hardest.

"Trucks are easier," mentioned Manny Manalastas, "because they have power steering."[73] Rodel Hernando, who has been driving a jeepney for ten years, mentioned the tight-shifting gears of the jeep, and how the car was far easier to drive.[74] Rolando Caoagas found the heat and exposure to the elements, such as the wind, particularly tiring and difficult.[75] Nevertheless, they continue to drive the jeepney because it provides them the freedom to work for themselves. One writer noted that the evolution of the jeepney was similar to that of comic strips in the Philippines; both started after World War II, and both provided some form of escape:[76] in this case, *escape from* an oppressive economic structure as well as *escape toward* an imagined reality. The driver was the king of a moving domain in a country marked by poverty.

The jeepney driver works for no one: he either owns his own jeepney or pays what in jeepney culture is called a "boundary," the daily minimum amount the driver needs to break even, the same concept as a rental fee. Once the driver "crosses" this boundary, the general allocation of earnings tends to follow a 40/60 rule; the driver keeps 40% of the total earnings while 60% goes to the owner of the jeepney. Apart from this informal financial arrangement, there has never been an overall governance structure that regulates jeepneys, just as the *calesas* were unregulated. Each driver was basically a free agent and hence king of his own jeepney domain.

The self-determination of jeepney drivers allowed them to conduct business as though, some would say, they owned the streets. Similar to the *calesas*, jeepneys would stop anywhere to pick up or drop off passengers. A prospective passenger would simply stand by the side of the road and flag down a jeepney. Getting off the jeepney would simply require one to say "Para," meaning "stop," a word used only in the context of a moving vehicle.

While some areas in Manila began to prohibit the frequent stopping and starting of vehicles along their roads, the majority of jeepneys still follow this *calesa* style of doing business. A Filipino writer stated, "Jeepney drivers stop every few meters, even in the middle of the road, or in corners, with no warning to, or consideration for, others."[77] Jeepney drivers do not like following rules, and many even stopped driving their vehicles in areas where traffic rules were starting to be enforced. Rolando Caoagas, for instance, drove jeepneys for twenty-six years in the busy profitable areas of Manila but moved to an upcountry province in the last six years because in the rural areas, he said, "Awan agtiltiliw," or nobody catches you [no enforcers].[78]

However, this aggressive driving contrasts with the religious elements found inside the jeep. Many observers have noted the "homey" look of the jeepney's interior, which often includes an altar propped up on the dashboard.[79] Michael Manalastas described having curtains, buying fresh flowers, and polishing the "God Bless Our Street" sign in his jeepney.[80] Thus, while the exterior was a manifestation of the festive flair of the *calesa*, the interior of the jeepney provided passengers a feeling of being a guest in the driver's home.

CONVERSION OF INSTITUTIONS: FROM MACHINELIKE TO HOMELIKE EFFECT, FILIPINO-STYLE

The jeepney represents various aspects of Filipino life.[81] One aspect often noted by many tourists is Filipino hospitality.[82] When someone visits a home, family members always provide some type of refreshment, even if it means sending someone off to the store through the back door. If someone arrives while the family is eating at the table, an extra place is immediately laid, and everyone adjusts their rice and fish intake to provide for the guest. This habit of accommodating guests in one's home extends to the habit of jeepney drivers accommodating anyone who flags them down, even if their vehicle is full. For instance, a six-passenger vehicle would be stretched to fit seven or eight, just as a Filipino family would accommodate additional guests regardless of its limitations.

One could argue that the driver is economically motivated to pack in as many passengers as possible. What is interesting in this phenomenon is the tolerance of passengers for such practices. Having ridden jeepneys for almost twenty years, I have never heard anyone complain about being packed in like sardines. It seems to be an accepted and understood practice to accommodate anyone out in the street trying to reach a destination, particularly at night. Many writers characterize this practice of trying to fit everyone into a vehicle as consistent with Filipino family values.[83]

Indeed, home and church are the two institutions that dominate a Filipino's life, and both are clearly evident in the jeepney's highly personalized embellishments. "The driver thinks of his jeepney, not as a simple vehicle with which he makes a living, but as an extension of his home, his church, his pride, his fears, his very self."[84] The names of his individual family members and relatives—and sometimes the driver's entire genealogy—are often painted all over the jeepney.[85] Every effort is made to provide the vehicle with as homelike an atmosphere as possible, particularly since the driver spends most of his day driving his vehicle along various routes for as many as ten to fifteen hours per day.

The addition of homelike touches, such as curtains along the windows of the jeepney and various knickknacks along the dashboard, gives the effect of being in someone's living room. Various Christian religious icons and images occupy the dashboard alongside the curios. It is common to find altars in many Filipino homes. Just as at home, many jeepney drivers dutifully buy fresh garlands of flowers to adorn the altar they have propped by their dashboard.[86] The religious icons were believed to provide protection from traffic, and in this sense, Christian figures became intermingled with animistic practices and beliefs. Filipino syncretic religiosity has been described by many observers as one of the dominant themes in many jeepney decorations.[87]

Thus, unlike in the United States where the shift from machinelike to coachlike effect involved passenger comfort, luxury, and convenience, the shift in the Philippine case was more a matter of incorporating into the jeepney the various institutions that comprise Filipino life. Riding in the jeepney was akin to being inside a home—albeit a modest one. Jeepney displays were often criticized by the middle class as gaudy and vulgar, in other words, *bakya*, which literally means "wooden clogs," the type of shoes associated with the lower class particularly in the rural areas.[88]

While the garishness of the heyday of the decorated jeeps has recently been tempered with more subdued decorations, the horse emblem persists, despite the demise of other decorations. A famous art critic, in a conversation with a well-known Filipino painter, observed the decline of the jeepney's festive air:

> In conversation Legaspi [a famous Filipino painter] agrees with the growing perception that, despite its exotic appeal to foreign tourists and local connoisseurs of nostalgia, this "motorized version of the *calesa*" is a sad reminder of our country's economic plight, which continues to get worse.[89]

Jeepneys appear to have been decreasing in significant numbers over the past several years. Sarao and Francisco went out of business many years ago and were not replaced by any new entrants. The names Sarao and Francisco, once prominently displayed on many jeepney creations, are now rarely seen. Nevertheless, one continues to see garishly decorated jeepneys, although they are a dying breed. While Sarao and Francisco manufactured locally made jeeps with engines imported from Japan, jeepneys began to be replaced by surplus air-conditioned vans from Korea and Japan.

Thus, while the initial push for the use of wartime jeeps was largely dominated by various foreign and local economic forces, the eventual commercialization of what would become known as the jeepney began at the grassroots level. The auto calesa that led to the jeepney became embodied in the chrome horse, which continues to take center stage on the hood of the jeepney, even after other decorations have waned in popularity.

However, when the last horse chrome falls, there is less motivation to replace it as cars and traffic congestion drive *calesas* out of city streets. Roger Abalos, a jeepney driver for fifteen years, did not bother to maintain the upkeep of his decorations when the last horse chrome fell off.[90] Similarly, when the horse chrome on John Corpuz's jeepney came off, he did not bother replacing it.[91] Diosdado Manalastas, who has driven a jeepney for eight years, scoffs at the entire idea of horse chrome and other jeepney decorations.[92]

Although Diosdado Manalastas is the exception among the fifteen drivers interviewed, his view nonetheless represents the weakened influence of horse-drawn carriages in modern transport. Motorization in the last decades had virtually erased the horse from modern memory. *Calesas* still meander along some street areas in Manila as well as in a few rural areas, but like the jeepney they are a dying breed. The jeepney, a mass-produced yet customized public vehicle that came to represent the fusion of the machinelike wartime jeep with the force of local *calesa* tradition, also appears to be facing its twilight years.

CONCLUSION

This chapter has shown the significant role the horse tradition played in domesticating motor power in Philippine society. The jeep arrived in a country made desolate by war. Yet the discarded surplus army vehicle rose to become a custom-built, elaborately ornamented passenger vehicle called the jeepney. Preexisting practices of communal riding, method of payment, and extravagant regal decorations of the *calesa* days transformed the military artifact into a popular mainstream transport of everyday life. Jeepneys became the mechanical equivalent of the ubiquitous *calesas* in spirit, look, and feel.

The legacy of the *calesa* practices persists in the form of a horse emblem situated on top of the hood. The horse emblem embodied the many threads that connected the past horse culture with its present symbolic mechanized form. From the driver's standpoint, the jeepney replicated the socioeconomic structure of the *calesa* and allowed freedom of self-expression through an automotive medium. The *calesa* brought a sense of belonging to the driver, whose movable dominion incorporated many symbols of local traditions. At the same time, the horse emblem provided means of escape to an imagined reality.

From the passenger's standpoint, the same sardine-packed communal riding and payment system from the *calesa* days persists. The honor system of payment and the practice of passing fares from one passenger to the next continue today. The hospitality practiced in Filipino homes came to be expected in the riding experience. The driver finds a way to accommodate everyone. The homelike interior décor of the jeep turned the passenger into a guest rather than a customer.

The notion of comfort in automotive design in jeepneys did not originate with women, as some gender analysts have argued to be the case in the U.S. setting. Historically, virtually all jeepney drivers have been men; they conceived the idea of transforming the jeepney into an extension of their homes. Automotive comfort, as Scharff argues, is a universally appreciated value, although in the case of the Philippines, it does not seem to have been motivated by gender-related factors alone.

From an artifactual perspective, the regal décor of the *calesa* was simply transferred to the jeepney. The *calesa* background of the jeepney pioneers came into play in the transformation of the jeep's technological form and spirit. The jeepney thus was purposely designed to exhibit a forced likeness with the *calesa*, just as early U.S. automotive pioneers tried to do with horseless carriages. However, there was a more seamless transfer of user practices from the *calesa* to the jeepney, perhaps because the concept of a self-propelled vehicle was not as controversial by that time or perhaps because the *calesa* influence pervaded the jeep thoroughly.

The complexity of the jeepney represented the convergence of Filipino institutions that transformed a foreign artifact into a usable representation of the *calesa*. This convergence, partly motivated by manufacturers and drivers, facilitated the acceptance of a foreign device into the everyday life of the Philippines. The case of the jeepney shows that persistence in practices permeated the form and functionality of the jeep, allowing a peripheral military object to become a mainstream yet sublime device of everyday life.

ENDNOTES

1. I searched the University of the Philippines and Ateneo de Manila University libraries for sources. I also used the Cornell University library system, known for its Southeast Asian studies collection (Kroch Library).
2. Having lived in the Philippines for more than twenty years, I found people to be extremely uncomfortable or embarrassed when being taped during interviews.
3. United States, Congress, House, *Subcommittee of the Committee on Public Buildings and Grounds, Hearings, Disposition of Surplus Property, Seventy-Eighth Congress, Second Session* (Washington, DC: GPO, 1943), 137.
4. John Anderson Miller, *Fares, Please! A Popular History of Trolleys, Horsecars, Streetcars, Buses, Elevateds, and Subways* (New York: Dover Publications, Inc., 1960), 147.
5. Teodoro A. Agoncillo, *History of the Filipino People* (Quezon City, Philippines: Garotech Publishing, 1990), 247.
6. Mary Lee Stubbs, Stanley Russell Connor, and United States Dept. of the Army, Office of Military History, *Armor-Cavalry Part II*, Army Lineage Series (Washington, DC: U.S. GPO, 1969), 29.
7. Agoncillo, *History of the Filipino People*, 219.
8. United States, Philippine Commission, 1899–1900, *Report of the Philippine Commission to the President, 1900*, vol. 1 [hereafter cited as *Report of the Philippine Commission to the President, 1900*] (Washington, DC: GPO, 1900), 174.
9. No insurgents were held because prisoners were shot.
10. *Report of the Philippine Commission to the President, 1900*, vol. 2 (Washington, DC: GPO, 1900), 71.
11. Ibid.
12. Ibid., 75.
13. Ibid.
14. *Report of the Philippine Commission to the President, 1900*, vol. 4 (Washington, DC: GPO, 1901), 27.
15. *Report of the Philippine Commission to the President, 1900*, vol. 3 (Washington, DC: GPO, 1901), 40.
16. Joseph P. McCallus, *American Exiles in the Philippines, 1941–1996* (Quezon City, Philippines: New Day Publishers, 1999), 42.
17. Agoncillo, *History of the Filipino People*, 400.

18. McCallus, *American Exiles in the Philippines, 1941–1996*, 63.
19. Trudl Dubsky Zipper and Herbert Zipper, *Manila 1944–45: As Trudl Saw It: Watercolors of Trudl Dubsky Zipper* (Santa Monica: Crossroads School in cooperation with Herbert Zipper, 1994), 11.
20. Ibid.
21. Walter Phelps Hall, *Iron Out of Calvary: An Interpretative History of the Second World War* (New York: D. Appleton-Century Company, Inc., 1946), 352.
22. McCallus, *American Exiles in the Philippines, 1941–1996*, 66.
23. Ibid., 66–67.
24. Zipper and Zipper, *Manila 1944–45: As Trudl Saw It*, 70.
25. The last Japanese stronghold in Manila was eliminated on March 4, 1945. See Zipper and Zipper, *Manila 1944–45: As Trudl Saw It*, 54.
26. *Motor Vehicle Law* (Manila: Bureau of Printing, 1955), 51.
27. Ibid.
28. Zipper and Zipper, *Manila 1944–45: As Trudl Saw It*, 56.
29. These early jeeps were most likely left behind by the United States in 1942 when it withdrew from the Philippines as Japanese forces swept the nation.
30. Major Arthur K. Whitehead, "With the 26th Cavalry (P.S.) in the Philippines," *The Cavalry Journal* 53, no. 3 (May–June, 1944), 34.
31. *The Papers of George Catlett Marshall*, ed. Larry I. Bland, asst. ed. Sharon R. Ritenour, asst. ed. Clarence E. Wunderlin, Jr. (Baltimore, MD: Johns Hopkins University Press, 1986), 577.
32. Janusz Piekalkiewicz, *The Cavalry of World War II* [Pferd und Reiter im II. Weltkrieg] (New York: Stein and Day Publishers, 1980), 212.
33. Emmanuel Torres, *Jeepney* (Quezon City, Philippines: GCF Books, 1979), 16.
34. Anjie Blardony Ureta, "The Jeepney is Still 'King of the Road,'" *Philippine Free Press*, October 12, 1991, 38.
35. Jose C. Kwe, "How the LRT Holds Up to the Philippine Jeepney," *WHO* (Manila), January 10, 1982, 15.
36. Torres, *Jeepney*, 16.
37. McCallus, *American Exiles in the Philippines, 1941–1996*, 71–72.
38. Ibid. Also, José S. Arcilla, S. J., *Recent Philippine History, 1898–1960* (Manila: Ateneo de Manila University Office of Research and Publications, 1997), 180.
39. Torres, *Jeepney*, 15. Also in Lamberto E. Antonio, "Pambihirang Sasakyan, May Lulang Kasaysayan," *Observer* (Manila), January 3, 1982, 13; Valerio Nofuente, "Jeepney: Vehicle as Art," *International Popular Culture* no. 1, 1 (1980): 38–47 (Reprinted in *Rediscovery: Essays on Philippine*

Life and Culture, rev. ed, ed. by Cynthia Nograles Lumbera and Teresita Gimenez-Maceda [N.p.: National Book Store, Inc., 1983]); Nestor Torre, Jr., "Jeepney Soul," *The Philippines*, 1983, 48.

40. Torres, *Jeepney*, 36–37; Randy V. Urlanda, "Jeep-Making: Imus' Sunshine Industry," *Philippine Panorama*, September 21, 1997, 12; "The Jeepney— A Remarkable Vehicle of Philippine Folk Art," *Philippine Hotel*, 1971, 19; Raymond J. de Souza, "View from a Manila Jeepney," Center for Research and Communication (CRC) *Staff Memos* (Manila), 1995, no. 12, 1; Ureta, "The Jeepney is Still 'King of the Road,'" 38; Randy V. Urlanda, "Jeep and Jeepney Assemblers Cooperative," *Philippine Panorama*, April 24, 1994, 10.

41. Ray Cowdery and Merrill Madsen, *All-American Wonder: Information Regarding the History, Production, Features and the Restoration of Military Jeeps, 1941–1945* (Rogers, MN: Victory Publishing Limited, 1993), 43.

42. Torres, *Jeepney*, 40. Also in Bo Bigkislahi, "Jeepney: Cacophony, Color, Culture," *Sunday Inquirer Magazine* (Manila), March 29, 1992, 12.

43. Jarius Y. Bondoc, "The Jeepney Kings," *Observer* (Manila), January 3, 1982, 11–12.

44. Ibid., 12.

45. Ibid.

46. "The Prime Movers," *Ibon Facts and Figures* (Manila), March 15, 1985, 2.

47. Construction was underway at this time to build a Light Railway Transit (LRT), though its route was very limited and was not widely used. See Jose M. Lansang, Jr., "LRT is a Misnomer," *Mr. & Ms.* (Manila), April 12–18, 1985, 21.

48. Torres, *Jeepney*, 48; Bigkislahi, "Jeepney: Cacophony, Color, Culture," 12.

49. Torres, *Jeepney*, 46.

50. Ibid., 48.

51. Urlanda, "Jeep-Making: Imus' Sunshine Industry," 12.

52. Bondoc, "The Jeepney Kings," 12. Also in Torres, *Jeepney*, 48.

53. Bondoc, "The Jeepney Kings," 12.

54. "The Jeepney—A Remarkable Vehicle of Philippine Folk Art," 19.

55. Torres, *Jeepney*, 58.

56. Nofuente, "Jeepney: Vehicle as Art," 42.

57. Ibid., 40.

58. George Basalla, *The Evolution of Technology* (Cambridge: Cambridge University Press, 1988), 106–107.

59. Urlanda, "Jeep-Making: Imus' Sunshine Industry," 12.

60. Ureta, "The Jeepney is Still 'King of the Road,'" 30.

61. Antonio, "Pambihirang Sasakyan, May Lulang Kasaysayan," 13.
62. Francoise Joaquin, "Facelift for the Faithful Jeepney," *Asia Magazine* (Manila), October 10–12, 1997, 6.
63. Interview with Dario Vega, October 29, 2005.
64. Interview with Rupino Prestosa, November 5, 2005.
65. Interview with Jojo Quines, October 29, 2005.
66. Bigkislahi, "Jeepney: Cacophony, Color, Culture," 12; Also in Urlanda, "Jeep and Jeepney Assemblers Cooperative," 10.
67. It is the same Ilokos region mentioned in the Lopez testimony in 1899.
68. Interview with Alfredo Bodayong, October 29, 2005.
69. Almost all jeepney drivers are male. I have ridden in jeepneys extensively for more than twenty years, and all the drivers I have encountered have been male.
70. Torres, *Jeepney*, 58.
71. Interview with Rupino Prestosa, November 5, 2005.
72. Interview with Jojo Quines, October 29, 2005.
73. Interview with Manny Manalastas, November 6, 2005.
74. Interview with Rodel Hernando, November 14, 2005.
75. Interview with Rolando Caoagas, November 5, 2005.
76. Torres, *Jeepney*, 42. Also in Antonio, "Pambihirang Sasakyan, May Lulang Kasaysayan," 13.
77. Rene Saguisag, "There's a Jeepney in the Filipino Soul," *Mr. & Ms.* (Manila), October 30, 1984, 10.
78. Interview with Rolando Caoagas, November 5, 2005.
79. Nofuente, "Jeepney: Vehicle as Art," 44–45. Also in de Souza, "View from a Manila Jeepney," 3; Ureta, "The Jeepney is Still 'King of the Road,'" 39.
80. Interview with Michael Manalastas, November 6, 2005.
81. Torres, *Jeepney*, 58. Antonio, "Pambihirang Sasakyan, May Lulang Kasaysayan," 13.
82. Nofuente, "Jeepney: Vehicle as Art," 46.
83. Ibid.
84. Torre, "Jeepney Soul," 49. For pictures of jeepney interiors, see Torres, *Jeepney*.
85. Ureta, "The Jeepney is Still 'King of the Road,'" 39.
86. "The Jeepney—A Remarkable Vehicle of Philippine Folk Art," 19.
87. de Souza, "View from a Manila Jeepney," 3.
88. Rolando S. Tinio, "Romancing the Jeepney," *Man Magazine* (Manila), January 1993, 4. Also, Antonio, "Pambihirang Sasakyan, May Lulang Kasaysayan," 13.

89. Emmanuel Torres, "Numero Uno's Pamana: Legaspi Reinvents the Jeepney," *Sunday Globe Magazine* (Manila), September 27, 1988, 11.
90. Interview with Roger Abalos, November 14, 2005.
91. Interview with John Corpuz, November 6, 2005.
92. Interview with Diosdado Manalastas, October 29, 2005.

CHAPTER 5

WHEN NEW THINGS
WERE OLD

In each of the three cases in this study, preexisting equine culture provided conceptual and material resources for those who sought to transform the motorcar into a mainstream form of public transit. Although faster than a horse and capable of carrying far heavier loads, this new technology had to become old in order to be accepted. The motorcar began as an unknown—and, in two cases, a threatening—new device.

The first case study shows that when the motorcar appeared in U.S. public streets at the turn of the twentieth century, its most noticeable feature—its lack of a horse—collided with the prevailing concept of motion as a product of muscle power. The newfangled machinery's most advantageous feature—the fact that it went without a horse—was also its chief liability.

In the second case study, the civilian origins of the motorcar prior to World War II stood at odds with the military's requirements for speed, light weight, and cross-country mobility, each of which the horse fulfilled effectively. Mounted attacks were central to the U.S. Cavalry's combat strategy. Forced motorization during the interwar years threatened its foundation.

In the third case study, the motorcar was less threatening. The military jeep was, after all, associated with the liberation of the Philippines from Japanese control. Nevertheless, having arrived from a different land, the jeep was exotic and, like the American motorcar in the early twentieth century, a novelty; it went without a horse in a land dominated by muscle power.

The features that alienated users from the motorcar in each case—the horselessness of the horseless carriage, the lack of warhorse capabilities in commercial vehicles, and the jeep's association with foreign military arms—had to be suppressed in order for it to gain mainstream status. In a context dominated by the horse, the staging of the motorcar as its operational and functional equivalent provided for an alternative.

In the first case study, automobile manufacturers and early adopters *propitiously leveraged* the concepts, practices, and even the reputation of the horse in order to facilitate the diffusion of the motorcar. Asserting a likeness with the sentient horse (motorcars as being sane,[1] responsive to the will of the driver,[2] instantly obedient,[3] as sensitive and spirited as thoroughbred horses[4]) did not require a change of work routines and infrastructure. Motor power simply replaced muscle power, while systems and structures of the horse tradition largely remained. Despite their significantly different operational controls (steering wheel touted to work like a pair of reins)[5] and their lack of equine "intelligence," motorcars came to pose as horse carriages through the public articulations of editorial cartoons, magazine articles, and advertisements.

In the second case study, it was the *purposeful effort* of the U.S. Cavalry to preserve its combat methods that led to the materialization of the jeep. While the first case study looks at rhetoric and imagery, the second case study examines how the horse concept persisted through a physical form. As in the first case, potential users were deeply hostile to the motorcar. The cavalry saw the passing of the horse as a threat. Efforts to protect the cavalry horse ironically led to its demise. Motorization, through the "breeding" of an iron warhorse, successfully supplanted the horse. Soldiers treated their jeeps as sentient machines, a scenario that early-twentieth-century advertisements strained to conjure.

In the third case study, preexisting equine practices *spontaneously effected* a new automotive hybrid, the jeepney. After World War II the surplus army jeep was domesticated for civilian use. Its adaptability in the Philippines was facilitated, as in the first two cases, by its role as a mechanical surrogate for the horse. Elements from the *calesa*, the Philippine horse-drawn carriage, transferred seamlessly to this new automotive medium. The transfer of decorative motifs, symbolic icons, and user practices localized the abandoned military jeeps in the postwar Philippines. Although the lack of alternative transport may have motivated the vehicle's adoption, local transformation came about as a result of the personal and institutional expressions of Philippine manufacturers and users.

In all three cases, because it was positioned to work and, to some extent, to look like a horse-drawn carriage, the automobile did not directly challenge entrenched work and sociocultural structures; it merely supplanted the horse. In each case, the compulsion to negate the motorcar's novelty—to make a new thing "old"—promoted technological diffusion. The motorcar simply replaced the horse in its Latourian black box.

Elements from the preexisting horse culture were incorporated to aid the diffusion of the new motorcar in all three cases: terminologies, nomenclature, material designs, functionality, operational controls, and infrastructure associated with the horse were employed to legitimize the emerging motor-powered vehicle. The appropriation of equine-transport practices and capabilities allowed the emerging motorcar to compete with the seemingly indispensable horse. The motorcar, ironically, came to dominate the horse by virtue of its association with it.

While historical and sociological studies of technological innovation tend to place new technologies at center stage, this study shows that the *new* depended on the *old*. Past studies typically emphasize technological diffusion as the triumph of the new over the old. Historians such as Berger and Flink note in particular the automobile's technical and economic advantages over the horse as motivations for adoption. Berger argues for the motorcar's greater power, performance, and efficiency,[6] while Flink points to its superior cleanliness, safety, and reliability.[7] To Kline and Pinch, "the advantages of the car became all too clear-cut."[8]

However, what neither technology studies nor the history of technology has addressed is the use of horse culture to create readily understandable concepts for the newly emerging and unfamiliar self-propelling vehicle. Not only did its innovative features—its ability to haul much heavier loads and run longer—facilitate the motorcar's adoption, so did its association with what it sought to replace. Established interpretations of an old artifact in this case inspired the comprehension and eventual diffusion of a new and even controversial device.

The skeuomorphic elements that Basalla considers nonfunctional in new artifacts may, in fact, serve a crucial purpose—to facilitate their use by defusing the threat posed by their novelty and occasional notoriety. The coachlike trimmings of horseless carriages and the jeepney's horse chromes, in this sense, were as functional as their mechanical parts. Such details evoked ease and familiarity. While Schivelbusch argues that the coachlike compartment failed to comfort perplexed upper-class travelers, he admits that, to his knowledge, in Europe "there were no attempts to create a passenger car that would be compatible in its form with the modern technology of the railroad—i.e., one that would no longer have anything to do with the coach-driven compartment."[9]

In the case study of the jeep and the jeepney, the old elements of the horse culture infused its replacement to such a degree that they virtually effaced the newness of motor power. Horseless carriages, on the other hand, failed to inspire the same level of devotion despite the aid of advertisements and print media. Its lack of a horse, as its name suggests, contrasts dramatically with the centrality of the horse in the jeep's functionality and the auto calesa's form. The auto calesa, a "horse-drawn carriage that happens to be automated," resembled the muscle power it replaced far more closely than its foreign, albeit motorized, source. Perhaps the extent to which the old pervades the new influences the success of domesticating radically new and controversial technologies.

What did provide admission for the motorcar in all three cases was its taming, or in Silverstone's parlance, its "invisibility."[10] This "invisibility" required the motorcar to appear to be as mundane and sensible as the horse by making *visible* in the new artifact what was *invisible* in the old—the safety, reliability, and ease of operation of the horse-drawn carriage, the ruggedness and cross-country capability of the war horse, and the flair and flamboyance of the *calesa*—all taken-for-granted features of the ubiquitous horse transport.

These equine articulations were set against the backdrop of many feminine themes. This study found that gasoline car advertisements directed at women began to appear as early as 1902, contrary to Scharff's claim that early automobile manufacturers initially marketed gasoline cars largely to men.[11] Even in the industry's early years, automobile manufacturers sold cars to anyone who could afford them. Reminiscences of early manufacturers evidence a determined effort to market to all potential *mainstream* users, particularly to women, who often appear as primary drivers in many early advertisements. Scharff's assumption that manufacturers delayed the diffusion of the gasoline car as a result of their mistaken, gendered worldview accords them too much discretion. After all, in a rising but still uncertain automotive industry, manufacturers were susceptible to the preferences of even small groups of customers.

Associating values, devices, and things with a certain gender appears to be a function of context. In the Philippines, automotive comfort was initiated by an all-male group of manufacturers and drivers, whereas Scharff posits that comfort in terms of ease of use was initially required by women in the United States. Historical evidence shows that early U.S. automobile manufacturers marketed clunky gasoline cars, not just nice little electric cars, to women across a broad economic spectrum. Many women used large gasoline cars even in the industry's early years.

The nascent automotive industry, supposedly a male dominion, depended on women's labor and patronage for technical skills and economic survival. An observer in 1898 noted that women were more enthusiastic than men about this new means of transport: women were the "most ardent promoters" of the motorcar and "created the first paying demand for a clean, speedy and reliable machine."[12] At the same time, women also played a role in the production of early automobiles. Women were described to perform the "better class of work" in almost every automobile plant, particularly for tops and upholstery.[13]

Historical evidence also shows that the jeep, later associated with masculine fervor, had feminine origins. The specifications demanded by the military for an iron horse became physically possible with the help of a car designed specifically for women learning to drive. The second case shows that technological artifacts associated with a certain gender, or agenda for that matter, are not predetermined but may change over time and according to use.

Focusing on how controversial technologies present themselves as variations on the old may provide necessary insights into how, in Latour's parlance, enrollment is achieved. The new is packaged and associated with the goodwill earned by old technologies. One could argue that even things possess reputation; radically new and controversial technologies intended for public use associate with reputable artifacts to bring familiarity and recognition.

The strategy of using the physical forms, practices, and infrastructures of the proven workhorse helped mitigate the liabilities posed by the roisterous devil wagon. Perhaps these types of associations provide currency on how public consensus could be reached, particularly for controversial nascent technologies. Such appropriations may be particularly effective for those devices for which reassurances of predictability and consistency are critical to acceptance.

The transition from muscle to motor power was a matter of connecting old ways with new things, a reversal of what most user studies tend to emphasize: the finding of new uses in old things. A successful campaign for technological change, particularly for highly controversial artifacts, may well require presenting the new as the familiar. A 1902 illustration from *The Automobile Magazine*[14] (Figure 67) perhaps foresaw how the motorcar would evolve during the next decade: *omnia mutantur*, "the more things change, the more they stay the same." Indeed, such was the case with the devil wagon.

FIGURE **67.** "The More Things Change, the More They Stay the Same."

Source. The Automobile Magazine, July–December 1902 (Brooklyn, NY: Electric Car Society, n.d. [CD-ROM]); illustration originally appeared on page 587 of the July 1902 issue.

ENDNOTES

1. *Collier's,* May 23, 1908, 27.
2. *Life*, March 12, 1903, 3.
3. *McClure's Magazine,* June 1904.
4. *Life*, May 5, 1904, 399.
5. *Life*, September 7, 1905, 266 (inside front cover).
6. Michael L. Berger, *The Devil Wagon in God's Country* (Hamden, CT: Archon Books, 1979), 34.
7. James J. Flink, *The Car Culture* (Cambridge, MA: MIT Press, 1975), 35. Also in James J. Flink and American Council of Learned Societies, *The Automobile Age* (Cambridge, MA: MIT Press, 1988), 138.
8. Ronald R. Kline and Trevor Pinch, "Users as Agents of Technological Change: The Social Construction of the Automobile in the Rural United States," *Technology and Culture* 37, no 4 (1996): 773.
9. Wolfgang Schivelbusch, *The Railway Journey: The Industrialization of Time and Space in the 19th Century* (Berkeley: University of California Press, 1986), 784.
10. Roger Silverstone, *Television and Everyday Life* (London: Routledge, 1994), 98.
11. Virginia Scharff, *Taking the Wheel: Women and the Coming of the Motor Age* (New York: Free Press, 1991), 37.
12. Henri Dumay, "The Locomotion of the Future," *Collier's,* July 30, 1898, 22.
13. Thomas J. Fay, "Trend in Design and Fashion" [subsection "Female Labor is Now Being Utilized"], *The Automobile,* December 30, 1909, 1148.
14. *The Automobile Magazine* is a different publication from *The Automobile.*

BIBLIOGRAPHY

Note: Because of the format of some of the archival sources, inclusive page numbers are unavailable for a few references. These cases are denoted by an asterisk after the cited page number(s).

Ackerson, Robert C. *Jeep: The 50 Year History*. A Foulis Motoring Book. Sparkford, Nr. Yeovil, Somerset, U.K.; Newbury Park, CA: Haynes, 1988.

Agoncillo, Teodoro A. *History of the Filipino People*. Quezon City, Philippines: Garotech Publishing, 1990.

Akrich, Madeline. "The De-scription of Technical Objects." In *Shaping Technology/Building Society*, edited by W. Bijker and J. Law, 205–24, 1992.

"All Pittsburgh Aroused." *The Automobile*, August 10, 1905, 174.

Allen, Terry de la Mesa. *Reconnaissance by Horse Cavalry Regiments and Smaller Units*. Harrisburg, PA: Military Service Pub. Co, 1939.

American Motor Corporation. *Annual Report*. 1971, 1973, 1976.

American Newspaper Annual and Directory. Philadelphia: N. W. Ayer & Son, 1924, 1930, 1935.

Antonio, Lamberto E. "Pambihirang Sasakyan, May Lulang Kasaysayan." *Observer* (Manila), January 3, 1982, 13.

Arcilla, José S., S. J. *Recent Philippine History, 1898–1960*. Manila: Ateneo de Manila University Office of Research and Publications, 1997.

Armored Cavalry Journal. 1946, 1947.

Askew, Mark. *Rare WW2 Jeep Photo Archive, 1940 to 1945*. Doncaster, U.K.: Jeep Promotions Ltd., 2001.

Atlantic Monthly. 1901.

"An Authority." "Automobile Fashions for Women: Newest French and American." *The Automobile and Motor Review*, November 1, 1902, 12.

"An Automobile Aid to Good Roads." *The Automobile*, April 8, 1905, 455–56 (cover).

The Automobile. 1899, 1903, 1905–1909.

The Automobile Magazine, July–December 1902. Brooklyn, NY: Electric Car Company, n.d. CD-ROM.

The Automobile and Motor Review. 1902.

"Automobiles Again Displace Horses." *The Horseless Age*, April 10, 1912, 657.

Automobiles: Jeep. A&E Television Networks, 1996.

"Automobiles vs. Horse-drawn Vehicles." *The Automobile*, August 10, 1905, 172–173.

"Automobilist Sues Policeman." *The Automobile*, July 4, 1903, 20–21.

"Autos Sold Before the Horse Block." *The Automobile*, May 31, 1906, 879.

"Auto Women Shoppers." *The Automobile*, May 3, 1906, 734.

Bairnsfather, Bruce. *Jeeps & Jests*. New York: Putnam, 1943.

Baker, Ray Stannard. "The Automobile in Common Use." *McClure's Magazine*, July 1899, 195–208.

Baldwin, Nick, G. N. Georgano, Michael Sedgwick, and Brian Laban. *The World Guide to Automobile Manufacturers*. New York: Facts on File Publications, 1987.

Basalla, George. *The Evolution of Technology*. Cambridge: Cambridge University Press, 1988.

Bate, T. R. F., Brigadier-General. "Buying British Remounts in America." In Galtrey, *The Horse and the War*, 27–35.

Behrendt, Hans-Otto. *Rommel's Intelligence in the Desert Campaign 1941–1943* [Rommels Kenntnis vom Feind im Afrikafeldzug]. London: W. Kimber, 1985.

Berger, Michael L. *The Devil Wagon in God's Country*. Hamden, CT: Archon Books, 1979.

Bigkislahi, Bo. "Jeepney: Cacophony, Color, Culture." *Sunday Inquirer Magazine* (Manila), March 29, 1992, 12.

Bijker, Wiebe E. "Do Not Despair: There Is Life After Constructivism." *Science Technology and Human Values* 18, no. 1 (1993): 113–138.

———. *Of Bicycles, Bakelites, and Bulbs*. Cambridge, MA: MIT Press, 1995.

———. "Sociohistorical Technology Studies." In *Handbook of Science and Technology Studies*, edited by S. Jasanoff, G. E. Markle, J. C. Petersen, and T. Pinch, 229–256. London: Sage, 1995.

Bijker, Wiebe E., and John Law. *Shaping Technology/Building Society: Studies in Sociotechnical Change*. Inside Technology. 1st MIT Press pbk. ed. Cambridge, MA: MIT Press, 1994.

Bimber, Bruce. "Three Faces of Technological Determinism." In *Does Technology Drive History?*, edited by Merrit Roe Smith and Leo Marx, 79–100. Cambridge, MA: MIT Press, 1994.

Blanco, Richard L. *Rommel, the Desert Warrior: The Afrika Korps in World War II*. New York: J. Messner, 1982.

Bloor, David. "Polyhedra and the Abominations of Leviticus." *British Journal for the History of Science* 11 (1978): 245–272.

Blumenson, Martin, and George S. Patton. *The Patton Papers*. Boston: Houghton Mifflin, 1972.

Bondoc, Jarius Y. "The Jeepney Kings." *Observer* (Manila), January 3, 1982, 11–12.

Bonie, Jean Jacques Théophile. *The French Cavalry in 1870*. Translated by C. F. Thomson. In *Cavalry Studies from Two Great Wars, Comprising the French Cavalry in 1870, by Lieutenant-Colonel Bonie, the German Cavalry in the Battle of Vionville—Mars-La-Tour, by Major Kaehler. The Operations of the Cavalry in the Gettysburg Campaign, by Lieutenant-Colonel George B. Davis*, edited by Captain Arthur L. Wagner, 9–129. International Series. Vol. 2. Kansas City, MO: Hudson-Kimberly Pub. Co., 1896.

Borg, Kevin. "The 'Chauffeur Problem' in the Early Auto Era: Structuration Theory and the Users of Technology." *Technology and Culture* 40, no. 4 (1999): 797–832.

Bourdieu, Pierre. *The Logic of Practice*. Translated by Richard Nice. Stanford, CA: Stanford University Press, 1990.

———. *Outline of a Theory of Practice*. Translated by Richard Nice. Cambridge: Cambridge University Press, 1977.

"Boy Knows His Car." *The Automobile*, October 17, 1903, 412.

Brandt, G., Lieutenant General. "Why Is the Cavalry Still Necessary?" *The Cavalry Journal* 41, no. 171 (May–June 1932): 46–47.

Brown, Arch, and the editors of *Consumer Guide. Jeep. 60th Anniversary Since 1941: The Unstoppable Legend*. Lincolnwood, IL: Publications International, 2001.

Butterworth, W. E. *Soldiers on Horseback: The Story of the United States Cavalry*. 1st ed. New York: W. W. Norton, 1967.

Caldwell, George L., Captain. "A History of Cavalry Horses." *The Cavalry Journal* 37, no. 153 (October 1928): 543–557.

Camp, Walter. "The Automobile." *Collier's*, March 9, 1901, 21.

Carlin, Benjamin. *Half-Safe: Across the Atlantic by Jeep*. London: A. Deutsch, 1955.

"Carriage Builders Consider Autos." *The Automobile*, October 12, 1905, 413.

Carroll, John, and Garry Stuart. *Classic Jeeps: The Jeep from World War II to the Present Day*. Osceola, WI: MBI Pub. Co., 2000.

CarsDirect. "1993 Cadillac DeVille Prices, Reviews, and Specs, at CarsDirect. com." http://www.carsdirect.com/research/cadillac/deville/1993/base.

"Cars New and Second-Hand." *The Automobile*, August 3, 1905, 146–147.

"Cars Offered For the Season of 1905." *The Automobile*, January 14, 1905, 49–63.

Carter, William Harding, Major General. "Early History of American Cavalry." *The Cavalry Journal* 34, no. 138 (January 1925): 7–8.

Cary, Norman Miller Jr. "The Use of the Motor Vehicle in the United States Army, 1899–1939." PhD diss., University of Georgia, 1980.

"Casual Cut-outs." *Dress and Vanity Fair*, October 1913, 59, 108.

"Cavalry Affairs before Congress." *The Cavalry Journal* 48, no. 211 (January–February 1939): 130–135.

"The Cavalry School of 1943." *The Cavalry Journal* 52, no. 1 (January–February, 1943): 85–89.

The Century. October 1898.

Chang, C. S. *The Japanese Auto Industry and the U.S. Market*. Praeger Studies in Select Basic Industries. New York: Praeger, 1981.

Chao, Sheau-yueh J. *The Japanese Automobile Industry: An Annotated Bibliography*. Bibliographies and Indexes in Economics and Economic History. Vol. 15. Westport, CT: Greenwood Press, 1994.

"Chief Characteristics of 1906 Models." *The Automobile*, January 11, 1906, 32–66.

Chrysler's Creation: 75th Anniversary. Flint, MI: McVey Marketing & Advertising, 2000.

Church, John. *Military Vehicles of World War 2*. Poole, Dorset, U.K.; New York: Blandford Press; Distributed in the United States by Sterling Pub. Co., 1982.

Churchill, Winston. *The Great Battles and Leaders of the Second World War: An Illustrated History* [Second World War]. Boston: Houghton Mifflin, 1995.

———. *Memoirs of the Second World War: An Abridgement of the Six Volumes of the Second World War*. Boston: Houghton Mifflin, 1959.

———. *Memoirs of the Second World War: An Abridgement of the Six Volumes of the Second World War with an Epilogue by the Author on the Postwar Years Written for this Volume*. Boston: Houghton Mifflin, 1990.

———. *The Second World War: The Grand Alliance*. 2nd ed. London: Cassell, 1950.

Clarke, R. M. *Jeep Collection No. 1*. Hong Kong: Brooklands Book Distribution Ltd., n.d.

"Club for Women Motorists in England." *The Automobile*, October 31, 1903, 451–52 (cover).

Clymer, Floyd. *Floyd Cramer's Historical Scrapbook. Motor Cars and News of 1899* [Historical scrapbook]. Los Angeles, CA, 1955.

———. *Henry's Wonderful Model T, 1908–1927*. New York: McGraw-Hill, 1955.

———. *Those Wonderful Old Automobiles*. New York: McGraw-Hill, 1953.

———. *Treasury of Early American Automobiles, 1877–1925*. New York: McGraw-Hill, 1950.

Cockburn, Cynthia. *Brothers: Male Dominance and Technological Change*. London: Pluto Press, 1983.

———. "The Material of Male Power." In *The Social Shaping of Technology*. Edited by Donald MacKenzie and Judy Wajcman, 177–198. Buckingham, U.K.: Open University Press, 1999.

Coffey, John W., and Williams College Museum of Art. *American Posters of World War One: Catalogue and Exhibition*. Williamstown, MA: Williams College Museum of Art, 1978.

Coldwell, Frederic L. *Preproduction Civilian Jeeps: 1944–1945 Models CJ-1 and CJ-2*. 1st ed. Denver, CO: Vintimage, 2001.

———. *Selling the All-American Wonder: The World War II Consumer Advertising of Willys-Overland Motors, Inc.* 1st American ed. Lakeville, MN: USM Inc., 1996.

Collier's. 1898, 1900, 1901, 1907–1909.

Collins, Harry M. *Artificial Experts: Social Knowledge and Intelligent Machines*. Cambridge, MA: MIT Press, 1990.

Colwell, M. Worth. "America's First Track Race." *The Horseless Age*, February 1, 1911, 272–274.

Commemorating 50 Years of the Jaunty Jeep. Flint, MI: McVey Marketing & Advertising Inc., 1998.

The Complete WW2 Military Jeep Manual. Andover, NJ: Portrayal Press, 1990–1993.

The Complete WW2 Military Jeep Manual. Hong Kong: Brooklands Books, Ltd., n.d.

"Considers Women Most Careful Drivers." *The Automobile*, September 5, 1907, 332.

Considine, John A., Lieutenant Colonel. "Sixth Cavalry-(Horse Mechanized) Fort Oglethorpe, Ga." *The Cavalry Journal* 50 (January–February 1941): 87–88.

"The Contributor's Club." *The Atlantic Monthly.* December 1901, 857–66.

Cowan, Ruth Schwartz. "The Industrial Revolution in the Home." In MacKenzie and Wajcman, *The Social Shaping of Technology*, 281–300.

———. *More Work for Mother*. New York: Basic Books, Inc., 1983.

Cowdery, Ray, and Merrill Madsen. *All-American Wonder: Information Regarding The History, Production, Features and the Restoration of Military Jeeps, 1941–1945*. Minneapolis, MN: Victory Publishing Limited, 1993.

Cowles, Virginia. *The Phantom Major: The Story of David Stirling and His Desert Command*. New York: Harper, 1958.

Crozier, William. *Ordnance and the World War: A Contribution to the History of American Preparedness*. New York: C. Scribner's Sons, 1920.

Cuntz, Hermann F. "The Automobile as a Feeder of Civilization." *The Automobile*, June 10, 1909, 952.

Daley, John L. S. "From Theory to Practice: Tanks, Doctrine, and the U.S. Army, 1916–1940." PhD diss., Kent State University, 1993.

"Dangerous Animals on the Streets." *The Automobile*, February 22, 1906, 434–435.

David, Paul. "Clio and the Economics of QWERTY." *American Economic Review* 75, no. 2 (1985): 332–337.

———. "Understanding the Economics of QWERTY: The Necessity of History." In *Economic History and the Modern Economist*, edited by W. N. Parker, 30–49. London: Basil Blackwell, 1986.

Davis, Charles Belmont. "The First Man Back." *Collier's*, November 7, 1908, 22–24.

Davis, Donald Finlay. *Conspicuous Production: Automobiles and Elites in Detroit, 1899–1933*. Philadelphia: Temple University Press, 1988.

Davis, J. M. "A Carriage Trade Viewpoint." Letter Box [letter to the editor]. *The Automobile*, May 18, 1905, 623–24 [including editor's response].

"Decay of Speed Laws and Ordinances." *The Automobile*, November 14, 1903, 499–501.

De Chasseloup-Laubat, Marquis. "Recent Progress of Automobilism in France." *The North American Review*, September 1899, 399–413.

Delaney, John. *Fighting the Desert Fox: Rommel's Campaigns in North Africa, April 1941 to August 1942*. London; New York: Arms and Armour; Distributed in the USA by Sterling Pub., 1998.

Denfeld, D. Colt, and M. Fry. *The Indestructible Jeep*. Ballantine's Illustrated History of the Violent Century. Weapons Book. Vol. 36. New York: Ballantine Books, 1973.

Derks, Scott, editor. *The Value of a Dollar: Prices and Incomes in the United States, 1860–1999*. Lakeville, CT: Grey House, 1999.

———. *The Value of a Dollar: Prices and Incomes in the United States, 1860–2004*. Millerton, NY: Grey House Publishing, 2004.

de Souza, Raymond J. "View from a Manila Jeepney." Center for Research and Communication (CRC) *Staff Memos* (Manila), 1995, no. 12, 1.

D'Este, Carlo. *Patton: A Genius for War*. 1st ed. New York: HarperCollins Publishers, 1995.

De Trafford, Humphrey Francis, and Walter Gilbey. *The Horses of the British Empire*. London: W. Southwood & Co., Limited, 1907.

DiNardo, R. L. *Mechanized Juggernaut or Military Anachronism? Horses and the German Army of World War II*. Contributions in Military Studies. Vol. 113. New York: Greenwood Press, 1991.

Dinsmore, Wayne. "What Every Horseman Should Know." *The Cavalry Journal* 34, no. 140 (July 1925): 292–295.

Domer, George Edward. "Good Things Did Come in Small Packages." *Automobile Quarterly* 14, no. 4 (1976): 405.

Dreyfus, Hubert L. *What Computers Can't Do: A Critique of Artificial Reason*. New York: Harper & Row, 1972.

Dumay, Henri. "The Locomotion of the Future." *Collier's*, July 30, 1898, 22–23.

Duncan, William Chandler. *U.S.-Japan Automobile Diplomacy: A Study in Economic Confrontation*. Cambridge, MA: Ballinger Pub. Co., 1973.

Duryea, Charles E. "As It Was in the Beginning." *The Automobile*, January 7, 1909, 47–48.

"Editor's Mail." *The Cavalry Journal* 50, no. 3 (May–June 1941): 46–47.

Edmunds, K. B., Lieutenant Colonel. "Tactics of a Mechanized Force: A Prophecy." *The Cavalry Journal* 39, no. 159 (July 1930): 410–17.

"Eighteen Months in Prison." *The Automobile*, August 17, 1905, 199.

Eisenhower, Dwight D. *Crusade in Europe*. New York: Doubleday and Company, Inc., 1948.

Encyclopedia of American Military History. General editor, Spencer C. Tucker; associate editors, David Coffey, John C. Fredriksen, and Justin D. Murphy. Vol. III, P to Z. New York: Facts on File, 2003.

Encyclopedia of the U.S. Census. Edited by Margo J. Anderson. Washington, DC: CQ Press, 2000.

An Ex-Cavalryman. "By Their Horses Ye Shall Know Them." *The Cavalry Journal* 33, no. 135 (April 1924): 199–200.

"Farm Horse Giving Way to Its Rival, the Auto." *The Automobile*, February 20, 1908, 246.

Fay, Thomas J. "Trend in Design and Fashion" [subsection "Female Labor is Now Being Utilized"]. *The Automobile*, December 30, 1909, 1148.

Fitch, George. "The Automobile." *Collier's*, September 19, 1908, 27–28.

Flink, James J. *America Adopts the Automobile, 1895–1910*. Cambridge, MA: MIT Press, 1970.

———. *The Car Culture*. Cambridge, MA: MIT Press, 1975.

Flink, James J., and American Council of Learned Societies. *The Automobile Age*. Cambridge, MA: MIT Press, 1988.

Fogel, Robert William. *The Escape from Hunger and Premature Death, 1700–2100: Europe, America, and the Third World*. Cambridge; New York: Cambridge University Press, 2004.

Forty, George. *The Armies of Rommel*. London; New York: Arms and Armour; Distributed in the USA by Sterling Pub., 1997.

Foster, Patrick. *Standard Catalog of Jeep, 1940–2003*. Iola, WI: Krause Publications, 2003.

Fowler, William. *Jeep Goes to War*. Philadelphia: Courage Books, 1993.

———. *SAS: Behind Enemy Lines: Covert Operations 1941 to the Present Day*. London: HarperCollins, 1997.

Fox, Charles Philip. *Working Horses: Looking Back 100 Years to America's Horse-Drawn Days: With 300 Historic Photographs*. 1st ed. Whitewater, WI: Heart Prairie Press, 1990.

Fox, Charles Philip, and Jean Van Dyke. *Horses in Harness*. Greendale, WI: Reiman Associates, 1987.

"Fox Hunting with an Automobile." *The Automobile*, November 30, 1905, 59 (cover).

Franz, Kathleen. *Tinkering: Consumers Reinvent the Early Automobile*. Philadelphia: University of Pennsylvania Press, 2005.

Fuller, J. F. C. *Armored Warfare, an Annotated Edition of Lectures on F.S.R. III (Operations between Mechanized Forces)*. Military Classics. 1st American ed. Harrisburg, PA: Military Service Pub. Co., 1943.

"Fundamentals of Cavalry Training Policy." Edited by Jerome W. Howe. *The Cavalry Journal* 30, no. 123 (April 1921): 180–186.

Furse, George Armand. *Military Transport*. London: Superintendence of Her Majesty's Stationery Office, 1882.

"Futile Deductions from Horse Statistics." *The Automobile*, March 29, 1906, 580.

Galtrey, Captain Sidney. *The Horse and the War*. London: Country Life, 1918.

Gardner, Bruce L. *American Agriculture in Twentieth Century: How It Flourished and What It Cost*. Cambridge, MA: Harvard University Press, 2002.

George C. Marshall Research Foundation, and Anthony R. Crawford. *Posters of World War I and World War II in the George C. Marshall Research Foundation*. Charlottesville, VA: University Press of Virginia, 1979.

George C. Marshall Research Foundation, and John N. Jacob. George C. Marshall Papers, 1932–1960: A Guide. Lexington, VA: George C. Marshall Research Foundation, 1987.

Ghosn, Carlos, and Philippe Ries. *Shift: Inside Nissan's Historic Revival*. New York: Currency Doubleday, 2003.

Giedon, Siegfried. *Mechanization Takes Command: A Contribution to Anonymous History*. New York: Oxford University Press, 1948.

Gillie, Mildred Hanson. *Forging the Thunderbolt: A History of the Development of the Armored Force*. Harrisburg, PA: Military Service Pub. Co., 1947.

Goode, Kenneth MacKarness. "Ten Years After: A Review of the Automobile Industry to the Present Day." *Collier's*, November 2, 1907, 12, 14.

Gordon, John W., and Theodore Ropp. *The Other Desert War: British Special Forces in North Africa, 1940–1943*. Contributions in Military Studies. Vol. 56. New York: Greenwood Press, 1987.

Gordon-Smith, Gordon, Captain. "The Role Played by the Serbian Cavalry in the World War." *The Cavalry Journal* 31, no. 128 (July 1922): 245.

Griess, Thomas E., and United States Military Academy. Dept. of History. *Atlas of the Second World War*. The West Point Military History Series. Wayne, NJ: Avery Pub. Group, 1985.

Grow, Robert W., Major. "Military Characteristics of Combat Vehicles." *The Cavalry Journal* 45, no. 6 (November–December 1936): 508–511.

Halberstadt, Hans. *Military Vehicles: From World War I to the Present*. New York: MetroBooks, 1998.

Hall, Walter Phelps. *Iron Out of Calvary: An Interpretative History of the Second World War*. New York and London: D. Appleton-Century Company, Inc., 1946.

Hamilton, H. G., First Lieutenant. "A Light Cross-Country Car." *The Cavalry Journal* 44, no. 189 (May–June 1935): 30.

Harbord, James G., Major General. "The Part of the Horse and the Mule in the National Defense." *The Cavalry Journal* 35, no. 143 (April 1926), 159–160*.

Harris, Townsend. *The Complete Journal of Townsend Harris*, introduction and notes by Mario Emilio Consenza. Tokyo: Charles E. Tuttle Company, 1930.

Hart, B. H. Liddell. *History of the Second World War*. New York: G.P. Putnam's Sons, 1971.

Hart, Victor. "Hunting by Automobile in England." *The Horseless Age*, February 28, 1912, 1.

Hasluck, Paul N., editor. *The Automobile: A Practical Treatise on the Construction of Modern Motor Cars: Steam, Petrol, Electric and Petrol-Electric*. Brooklyn, NY: Electric Car Society, 2003. CD-ROM.

Hawkins, H. S., Colonel. "The Importance of Modern Cavalry and Its Role as Affected by Developments in Airplane and Tank Warfare." *The Cavalry Journal* 35, no. 145 (October 1926): 489*.

Haynes, Elwood. "A Few Reminiscences of the Early Automobile." *The Horseless Age*, December 27, 1911, 957.

Hayward, Charles B. "How the Horse and It's [sic] Load Wear Out Roads." *The Automobile*, June 18, 1908, 843–846.

Henry, Guy V., Major General. "The Trend of Organization and Equipment of Cavalry in the Principal World Powers and Its Probable Role in Wars of the Near Future." *The Cavalry Journal* 41, no. 170 (March–April 1932): 5–9.

Herr, John K., Major General, and Edward S. Wallace. *The Story of the U.S. Cavalry, 1775–1942*. 1st ed. Boston: Little, Brown, 1953.

Hilgartner, Stephen. "The Social Construction of Risk Objects: Or, How to Pry Open Networks of Risk." In *Organizations, Uncertainties, and Risk*, edited by James F. Short, Jr., and Lee Clarke, 39–53. Boulder, CO: Westview Press, 1992.

Hill, Russell. *Desert War*. New York: Alfred A. Knopf, 1942.

Hino, Corporal Ashihei. *Wheat and Soldiers*. New York: Farrar & Rinehart, Inc., 1939.

Hiscox, Gardner Dexter, and Electric Car Society. *Horseless Vehicles, Automobiles, Motorcycles*. Brooklyn, NY: Electric Car Society, 2003. CD-ROM.

Hitchcock, Mrs. A. Sherman. "Woman at the Wheel." *The Automobile*, May 6, 1909, 753–754.

———. "Women as Drivers of Automobiles." *The Automobile*, April 19, 1906, 674.

Hogg, Ian V., and John S. Weeks. *The Illustrated Encyclopedia of Military Vehicles*. 1st American ed. Englewood Cliffs, NJ: Prentice-Hall, 1980.

Holmes, Torlief S. *Horse Soldiers in Blue*. Gaithersburg, MD: Butternut Press, 1985.

Homans, James E. *Self-Propelled Vehicles: A Practical Treatise on the Theory, Construction, Operation, Care, and Management of All Forms of Automobiles*. New York: Theo. Audel & Company, 1905.

Home, Archibald, and Diana Briscoe. *The Diary of a World War I Cavalry Officer*. Tunbridge Wells, Kent, U.K.: D. J. Costello, Ltd., 1985.

"Horse-Drawn Statistics That Indicate Auto's Growth." *The Automobile*, February 6, 1908, 189.

"Horse Firm Takes Auto Agency." *The Automobile*, April 1, 1909, 557.

The Horseless Age: The Automobile Trade Magazine. 1895, 1909–1912, 1918.

"The Horseless Carriage." About the World. *Scribner's Magazine*, March 1896, 393–394.

"Horses and Motors." *The Cavalry Journal* 45, no. 194 (March–April 1936): 105.

"The Horses Come Before Anything Else." *The Cavalry Journal* 37, no. 152 (July 1928): 415–418.

"How To Be Hideous." *The Automobile*, July 11, 1903, 29.

Howze, Hamilton H. *A Cavalryman's Story: Memoirs of a Twentieth-Century Army General*. Washington, DC: Smithsonian Institution Press, 1996.

Huerbana, Ricardo R. *Philippine Folk Beliefs and Customs File: Use, Organization, Source, and Method of Analysis & Classification*. Manila: Xavier University, 1965.

Hughes, Thomas P. "The Evolution of Large Technological Systems." In *The Social Construction of Technological Systems*, edited by Wiebe E. Bijker, Thomas P. Hughes, and Trevor J. Pinch, 51–82. Cambridge, MA: MIT Press, 1987.

Hughes, Thomas Parke. *Networks of Power: Electrification in Western Society, 1880–1930*. Baltimore, MD: Johns Hopkins University Press, 1983.

Humphreys, Phebe Westcott. "An Automobile Vacation on $1.60 a Day." *The Ladies Home Journal*, July 1906, 27.

Hunt, Frazier, and Robert Hunt. *Horses and Heroes: The Story of the Horse in America for 450 Years*. New York: C. Scribner's Sons, 1949.

Hunt, Frazier, and Douglas MacArthur. *MacArthur and the War Against Japan*. New York: C. Scribner's Sons, 1944.

Hutton, F. R. "Technical Experience" [contribution to "Trade Outlook for the Coming Year"]. *The Automobile*, January 14, 1905, 42.

"Imperfect." *Life*, November 21, 1901, 415.

"In Touch with Market." *The Automobile*, October 5, 1905, 370.

"Jail Sentences for Speeders." *The Automobile*, June 8, 1905, 707.

James, Dorris Clayton. *The Years of MacArthur*. Boston: Houghton Mifflin, 1970.

"The Jeepney—A Remarkable Vehicle of Philippine Folk Art." *Philippine Hotel*, 1971, 19.

"Jeeps on the Farm." *Popular Mechanics*, January 1943, 46–47.

Jepsen, Stanley M. *The Coach Horse: Servant with Style*. South Brunswick, NJ: A. S. Barnes and Co., Inc., 1977.

———. *The Gentle Giants: The Story of Draft Horses*. South Brunswick, NJ: A. S. Barnes, 1971.

Jeudy, Jean Gabriel, and Marc Tararine. *The Jeep*. Warne's Transport Library. London: Warne, 1981.

Joaquin, Francoise. "Facelift for the Faithful Jeepney." *Asia Magazine*, October 10–12, 1997, 6.

Joerges, Bernward. "Do Politics Have Artifacts?" *Social Studies of Science* 29, no. 3 (1999): 411–431.

Johnson, David E. *Fast Tanks and Heavy Bombers: Innovation in the U.S. Army, 1917–1945*. Cornell Studies in Security Affairs. Ithaca, NY: Cornell University Press, 1998.

Jordan, Kathleen, and Michael Lynch. "The Sociology of a Genetic Engineering Technique: Ritual and Rationality in the Performance of the 'Plasmid Prep.'" In *The Right Tool for the Job*, edited by Adele Clarke and Joan Fujimura, 77–114. Princeton, NJ: Princeton University Press, 1992.

Josephy, Alvin. "How Do You Stay Alive?" In Smith, *The United States Marine Corps in World War II*, 814–817.

Kaempfer, Engelbert. *The History of Japan*. London: Thomas Woodward and Charles Davis, 1727.

Kennett, Lee B. *GI: The American Soldier in World War II*. 1st ed. New York: Charles Scribner's Sons, 1987.

Kenrick, Vivienne. *Horses in Japan*. London: J. A. Allen, 1964.

Kline, Ronald R. *Consumers in the Country: Technology and Social Change in Rural America*. Revisiting Rural America. Baltimore, MD: Johns Hopkins University Press, 2000.

Kline, Ronald R., and Trevor Pinch, "Users as Agents of Technological Change: The Social Construction of Automobile in the Rural United States." *Technology and Culture* 37, no. 4 (1996): 763–795.

Krarup, Marius C. "Influence of Carriage Styles on the Construction of Automobiles." *The Automobile*, April 11, 1903, 396–400.

Kriebel, Rainer, Bruce I. Gudmundsson, and United States War Dept. Military Intelligence Service. *Inside the Afrika Korps: The Crusader Battles, 1941–1942*. London; Mechanicsburg, PA: Greenhill Books; Stackpole Books, 1999.

Kühn, Volkmar. *Rommel in the Desert: Victories & Defeat of the Afrika-Korps, 1941–1943*. Schiffer Military History. [Mit Rommel in der Wüste.]. West Chester, PA: Schiffer Pub., 1991.

Kwe, Jose C. "How the LRT Holds Up to the Philippine Jeepney." *WHO* (Manila), January 10, 1982, 15.

Laird, Pamela Walker. "The Car Without a Single Weakness: Early Automobile Advertising." *Technology and Culture* 37, no. 4 (1996): 796–812.

Lakatos, Imre. *Proofs and Refutations: The Logic of Mathematical Discovery*. Edited by John Worrall and Elie Zahar. Cambridge: Cambridge University Press, 1976.

Lambert, Zita Elaine, and Robert John Wyatt. *Lord Austin: The Man*. London: Sidgwick & Jackson, 1968.

Landes, David S. *Revolution in Time: Clocks and the Making of the Modern World*. Cambridge, MA: Belknap Press of Harvard University Press, 1983.

Landis, Carole. *Four Jills in a Jeep*. New York: Random House, 1944.

Lansang, Jose M. Jr. "LRT is a Misnomer." *Mr. & Ms.*, April 12–18, 1985, 21.

Latour, Bruno. *Aramis, or, The Love of Technology*. Translated by Catherine Porter. Cambridge, MA: Harvard University Press, 1996.

Latour, Bruno (writing as Jim Johnson). "Mixing Humans and Nonhumans Together: The Sociology of a Door-Closer." *Social Problems* 35, no. 3 (1988): 298–310.

Latour, Bruno. *Science in Action*. Cambridge, MA: Harvard University Press, 1987.

Law, John. "Technology and Heterogeneous Engineering: The Case of the Portuguese Expansion." In *The Social Construction of Technological Systems: New Directions in the Sociology and History of Technology*, edited by W. E. Bijker, T. P. Hughes, and T. J. Pinch, 111–34. Cambridge, MA: MIT Press, 1987.

Life. 1898, 1901–1905, 1907, 1910, 1913, 1915, 1916, 1918.

Lininger, Clarence, Lieutenant Colonel. "Mobility, Fire Power, and Shock." *The Cavalry Journal* 34, no. 139 (April 1925): 178–181.

Lucas, Jim. "Occupation of the Russells." In Smith, *The United States Marine Corps in World War II*, 359–371.

Lynn, Thomas. "Father of the Jeep." *Journal of America's Military Past* 24, no. 3 (1997): 25–30.

MacGlashan, Katrine. *Horseless Buggy*. Boston: Little, Brown, 1942.

MacKenzie, Donald A., and Judy Wajcman. *The Social Shaping of Technology*. 2nd ed. Buckingham, U.K.; Philadelphia: Open University Press, 1999.

"Man, Horse, Automobile, and the Highways." *The Automobile*, January 9, 1908, 38–40.

Mariot, Chef D'Escadrons Breveté. "The Cavalry's Problem." *The Cavalry Journal* 43, no. 183 (May–June 1934): 11–15.

Marshall, George C., Larry I. Bland, and Sharon Ritenour Stevens. *The Papers of George Catlett Marshall*. Baltimore, MD: Johns Hopkins University Press, 1981.

Marshall, George C., Larry I. Bland, ed., Sharon R. Ritenour, asst. ed., Clarence E. Wunderlin, Jr., asst. ed. *The Papers of George Catlett Marshall*. Baltimore, MD: Johns Hopkins University Press, 1986.

Marshall, George C., Forrest C. Pogue, and Larry I. Bland. *George C. Marshall Interviews and Reminiscences for Forrest C. Pogue*. Rev. ed. Lexington, VA: George C. Marshall Research Foundation, 1991.

"Massachusetts Automobile Law." *The Automobile and Motor Review*, June 14, 1902, 20–21.

"Massachusetts Automobile Law Signed by Governor Bates." *The Automobile*, July 4, 1903, 21–22.

Mauldin, Bill. *Back Home*. New York: W. Sloan Associates, 1947.

———. *Bill Mauldin's Army*. New York: Sloane, 1951.

———. *Let's Declare Ourselves Winners—and Get the Hell Out*. Novato, CA: Presidio, 1985.

———. *Up Front: Text and Pictures*. New York: H. Holt and Company, 1945.

———. *Up Front*. Cleveland: The World Publishing Company, 1946.

McCallus, Joseph P. *American Exiles in the Philippines, 1941–1996*. Quezon City, Philippines: New Day Publishers, 1999.

McClure's Magazine, 1896, 1899, 1901, 1902, 1904, 1905.

McGaw, Judith A. "Why Feminine Technologies Matter." In *Gender & Technology*, edited by Nina E. Lerman, Ruth Oldenziel, and Arwen P. Mohun, 13–36. Baltimore, MD: Johns Hopkins University Press, 2003.

McGroarty, John S. "The Valley of Surprise." *The West Coast Magazine*, June 1911, 266–267.

McGuire, E. C., Major. "Armored Cars in the Cavalry Maneuvers." *The Cavalry Journal* 39, no. 160 (July 1930): 386–399.

McHarg, A. V. A. "A Story of the Cry of 'Get A Horse.'" *The Automobile*, January 16, 1908, 79.

McKee, Oliver, Jr. "With the 'Cavalree' at Fort Riley." *The Cavalry Journal* 34, no. 138 (January 1925): 71–76.

McShane, Clay. *Down the Asphalt Path: The Automobile and the American City*. New York: Columbia University Press, 1994.

McShane, Clay, and Joel A. Tarr, "The Centrality of the Horse in the Nineteenth-Century American City." In *The Making of Urban America*, edited by Raymond A. Mohl, 105–130. Wilmington, DE: Scholarly Resources Inc., 1997.

Mechanics Magazine. January 1, 1834, 16.

"Mechanized Force Becomes Cavalry." *The Cavalry Journal* 40, no. 165 (May–June 1931): 5–6.

Meneely, A. Howard. *The War Department, 1861: A Study in Mobilization and Administration.* Studies in History, Economics and Public Law. Vol. 300. New York: Columbia University Press, 1928.

Miller, John Anderson. *Fares, Please! A Popular History of Trolleys, Horsecars, Streetcars, Buses, Elevateds, and Subways.* New York: Dover Publications, Inc., 1960.

Miller, Lee Graham. *The Story of Ernie Pyle.* New York: Viking Press, 1950.

Mitcham, Samuel W. *Rommel's Desert War: The Life and Death of the Africa Korps.* New York: Stein and Day, 1982.

Moffett, Cleveland. "The Edge of the Future." *McClure's Magazine,* July 1896, 153–156.

Mohl, Raymond A., editor. *The Making of Urban America.* 2nd ed. Wilmington, DE: Scholarly Resources, 1997.

Moore, John, Sir. *Our Servant the Horse: An Appreciation of the Part Played by Animals during the War, 1914–1918.* London: H. & W. Brown, 1931.

Morison, Elting Elmore. *Men, Machines, and Modern Times.* Cambridge, MA: MIT Press, 1966.

Morrell, De Witt C. "Reckless Employers" [in "Comments and Queries of Readers"]. *The Horseless Age,* December 8, 1909, 652.

Morrison, George. "The New Epoch and the Currency." *The North American Review,* February 1897, 139–150.

Morse, Albert Whipple, Jr., Major. "Stables, New Type." *The Cavalry Journal* 50, no. 3 (May–June 1941): 75.

Moseley, George Van Horn, Major General. "Industry and National Defense." *The Cavalry Journal* 40, no. 162 (January 1931): 17–19.

"Motor Car Law." *Motor*, June 1909, 40–45.

"Motor Carriages" [from *The Horseless Carriages* by Prof. John Trow-bridge]. *The Living Age Company*, April 10, 1897, 131–132.

"Motor Life Stirring In the Iron City." *The Automobile*, December 5, 1903, 589–590.

"Motor Trucks Cheaper Than Horses." *The Horseless Age*, April 12, 1911, 632.

Motor Vehicle Law. Manila: Bureau of Printing, 1955.

Mott, Frank Luther. *A History of American Magazines 1885–1905*. Cambridge, MA: Harvard University Press, 1957.

Mottistone, John Edward Bernard Seely, Baron. *My Horse, Warrior*. London: Hodder & Stoughtor, Ltd., 1934.

Munro, Bill. *Jeep: From Bantam to Wrangler*. Marlborough, Wiltshire, U.K.: Crowood Press Ltd., 2000.

Munsey, Frank A. "The Automobile in America." *The Automobile*, February 1, 1906, 313–315.

Musselman, M. M. *Get a Horse! The Story of the Automobile in America*. Philadelphia: J. B. Lippincott Company, 1950.

Nabholtz, Lawrence. *The Military Jeep, Model MB-GPW: An Illustrated Guide to its Features and Evolution, 1941–1945*. Linden, TX: L. Nabholtz, 1996.

Nash, Lyman M. "The True History of the Ugly." *The American Legion Magazine*, February 1967. In Richards, *Military Jeeps 1941–1945*, 48–53.

Nason, Leonard, Captain. "Horse and Machine." *The Cavalry Journal* 38, no. 155 (April 1929): 192–195.

"Necessity for Horsed Cavalry Under Modern Conditions: Extract from the Recent Hearings Before the Subcommittee of the Committee on Appropriations, House of Representatives, on the War Department

Appropriation Bill, 1938." *The Cavalry Journal* 46, no. 201 (May–June 1937): 251.

"New Street Traffic Ordinance for New York." *The Automobile*, January 31, 1903, 154.

Newman, Bernard. *The Cavalry Goes Through!* New York: H. Holt and Company, 1930.

"The Next Exhibition of Automobiles" [in French; adapted from *Journal Amusant*]. *Life*, August 4, 1898, 99.

Noble, David F. *Forces of Production: A Social History of Industrial Automation*. New York: Alfred A. Knopf, 1984.

———. "Social Choice in Machine Design: The Case of Automatically Controlled Machine Tools." In MacKenzie and Wajcman, *The Social Shaping of Technology*, 161–176.

Nofuente, Valerio. "Jeepney: Vehicle as Art." *International Popular Culture* no. 1, 1 (1980): 38–47. Reprinted in *Rediscovery: Essays on Philippine Life and Culture*. Rev. ed. Ed. by Cynthia Nograles Lumbera and Teresita Gimenez-Maceda. N.p.: National Book Store, Inc., 1983.

The North American Review. 1897, 1899, 1900.

Ogburn, William F. "How Technology Changes Society." *Annals of the American Academy of Political and Social Science* 249, no. 1 (January 1947): 81–88.

Ogburn, William F. *On Culture and Social Change: Selected Papers*. The Heritage of Sociology. Chicago: University of Chicago Press, 1964.

———. *Social Change*. New York: Viking Press, 1922.

Ohmann, Richard. *Selling Culture: Magazines, Markets, and Class at the Turn of the Century*. London: Verso, 1996.

Oldenziel, Ruth. "Why Masculine Technologies Matter." In *Gender & Technology*, edited by Nina E. Lerman, Ruth Oldenziel, and Arwen P. Mohun, 37–71. Baltimore, MD: Johns Hopkins University Press, 2003.

O'Malley, T. J., and Ray Hutchins. *Military Transport: Trucks & Transporters*. Greenhill Military Manuals. London: Greenhill Books, 1995.

"One of the Faithful." "Faith in and a Doctrine for the Cavalry Service." *The Cavalry Journal* 36, no. 147 (April 1927): 227*.

Orton, Ernest Frederick. *Cavalry Taught by Experience: A Forecast of Cavalry under Modern War Conditions, "by Nortrofe."* London: Hugh Rees, 1910.

Orton, J. W. *The Story of Semiconductors*. Oxford, U.K.: Oxford University Press, 2004.

Oudshoorn, Nelly, and Trevor Pinch, ed. *How Users Matter: The Co-Construction of Users and Technologies*. Cambridge, MA: MIT Press, 2003.

Palmer, Bruce, Captain. "The Bantam in the Scout Car Platoon." *The Cavalry Journal* 50, no. 2 (March–April 1941): 89–92.

Parker, James, Brigadier General. "The Cavalryman and the Rifle." *The Cavalry Journal* 37, no. 152 (July 1928): 362–367.

Patton, George S., Major. "Armored Cars with Cavalry." *The Cavalry Journal* 33, no. 134 (January 1924): 5–10.

———. "Mechanization and Cavalry." *The Cavalry Journal* 39, no. 159 (April 1930): 234–240.

———. "Motorization and Mechanization in the Cavalry." *The Cavalry Journal* 39, no. 160 (July 1930): 331–348.

———. "What the War Did for Cavalry." *The Cavalry Journal* 31, no. 127 (April 1922): 165–172.

Payne, Robert. *The Marshall Story: A Biography of General George C. Marshall*. 1st ed. New York: Prentice-Hall, 1951.

Petterson, Daniel. "Algodones Dunes: Most Illegal Place in the World." *Earth First!*, July 31, 2000, 18.

Phillips, Albert E., Colonel. "The First Motorized Cavalry." *The Cavalry Journal* 43, no. 183 (May–June 1934): 10–11.

Piekalkiewicz, Janusz. *The Cavalry of World War II.* [Pferd und Reiter im II. Weltkrieg.] New York: Stein and Day, 1980.

Pimlott, John, and Alan Bullock. *The Historical Atlas of World War II.* Henry Holt Reference Book. 1st ed. New York: H. Holt, 1995.

Pinch, Trevor. "Giving Birth to New Users: How the Minimoog Was Sold to Rock and Roll." In *How Users Matter: The Co-Construction of Users and Technologies*, edited by Nelly Oudshoorn and Trevor Pinch, 247–270. Cambridge, MA: MIT Press, 2003.

———. "The Social Construction of Technology: A Review." In *Technological Change*, edited by Robert Fox, 17–36. Amsterdam: Harwood, 1996.

Pinch, Trevor, and Wiebe Bijker. "The Social Construction of Facts and Artifacts: Or How the Sociology of Science and the Sociology of Technology Might Benefit Each Other." In *The Social Construction of Technological Systems*, edited by Wiebe Bijker, Thomas Hughes and Trevor Pinch, 17–50. Cambridge, MA: MIT Press, 1989.

Pinch, Trevor, and Karin Bijsterveld. " 'Should One Applaud?' Breaches and Boundaries in the Reception of New Technology in Music." *Technology and Culture* 44, no. 3 (2003): 536–559.

Pinch, Trevor, and Frank Trocco. *Analog Days.* Cambridge, MA: Harvard University Press, 2002.

"Pleasure vs. Commercial Cars." *The Automobile*, April 12, 1906, 651.

Pogue, Forrest C. *Diaries of a WWII Combat Historian.* Lexington: University Press of Kentucky, 2001.

———. *George C. Marshall.* New York: Viking Press, 1963.

"Poor Lo Takes an Automobile Outing." *The Automobile*, May 18, 1905, 609 (cover).

"The Poor Man's Automobile." *The Horseless Age*, July 12, 1911, 49.

Postrel, Virginia I. *The Future and Its Enemies: The Growing Conflict over Creativity, Enterprise, and Progress*. New York: Simon & Schuster, 1999.

Pride, W. F., Lieutenant. "Principle of the Offensive." *The Cavalry Journal* 35, no. 142 (January 1926): 55*.

"The Prime Movers." *Ibon Facts and Figures*, March 15, 1985, 2.

"The Private Automobile Stable." *The Automobile and Motor Review*, July 5, 1902, 1–3.

Probst, Karl K., with Charles O. Probst. "One Summer in Butler—Bantam Builds the Jeep." *Automobile Quarterly* 14, no. 4 (1976): 431–438.

Public Employees for Environmental Responsibility. "BML Will Sign Flawed Algondones Dunes Off-Road Plan." News Release, March 24, 2005, http://www.peer.org/news/news_id.php?row_id=500.

Pyle, Ernie. *Brave Men*. New York: Grosset & Dunlap, 1945.

———. *G. I. Joe*. Vol. F[rench] 10. New York: Overseas Editions, Inc., 1944.

———. *Here is Your War*. New York: Henry Holt and Company, 1943.

———. *Here is Your War. Story of G. I. Joe*. A Forum motion picture ed. Cleveland: World Pub. Co., 1945.

———. *Last Chapter*. New York: H. Holt and Company, 1946.

Pyle, Ernie, and David Nichols. *Ernie's America: The Best of Ernie Pyle's 1930's Travel Dispatches*. New York: Random House, 1989.

Pyle, Ernie, and David Nichols, ed. *Ernie's War: The Best of Ernie Pyle's World War II Dispatches*. 1st ed. New York: Random House, 1986.

Rae, John B. "The Rationalization of Production." In *Technology in Western Civilization*, edited by Melvin Kranzberg and Carroll W. Pursell, Jr., 37–52. New York: Oxford University Press, 1967.

Ramsay, A. R. "Four Women and an Auto." *The Automobile*, June 24, 1909, 1044.

Randolph, James R., Major. "Mental Mobility." In *The Cavalry and Armor Heritage Series*, edited by Royce R. Taylor, Jr., 59–66. Fort Knox, KY: United States Armor Association, 1986. Reprinted from *The Cavalry Journal* 49, no. 217 (January–February 1940): 10*.

"Reckless Chauffeurs Arrested." *The Automobile*, January 21, 1905, 158–159.

Reed, David. *The Popular Magazine in Britain and the United States 1880–1960*. Toronto: University of Toronto Press, 1997.

Reilly, Henry J., Brigadier General. "Horse Cavalry and the Gas Engine's Children." *The Cavalry Journal* 49, no. 217 (January–February 1940): 2–9.

Richards, T. *Military Jeeps, 1941–1945*. Brooklands Road & Track Series. Bloomfield, NJ; Cobham, Surrey, U.K.: Portrayal Press; Distributed by Brooklands Book Distribution, 1985.

Rifkind, Herbert R. *The Jeep: Its Development & Procurement under the Quartermaster Corps, 1940–1943*. London: ISO Publications, 1943.

Risch, Erna. *Quartermaster Support of the Army: A History of the Corps, 1775–1939*. CMH Pub. Vol. 70-35. Washington, DC: Center of Military History, U.S. Army: For sale by the Supt. of Docs., U.S. GPO, 1989.

"Rollover: The Hidden History of the SUV." *Frontline*. February 21, 2002. http://www.pbs.org/wgbh/pages/frontline/shows/rollover/etc/script. html.

Rommel, Erwin, and Basil Henry Liddell Hart, editor. *The Rommel Papers*. With the assistance of Lucie-Maria Rommel, Manfred Rommel, and General Fritz Bayerlein, translated by Paul Findlay, 1st American ed. New York: Harcourt, Brace and Company, 1953.

Roos, Delmar Gerle. Delmar Gerle Roos papers, 1952–1953. [Album of photographs of Willys jeeps used as farm tractors with various farm implements. Also, blueprint from Willys-Overland Motors of an engine assembly and test reports of a modified engine.] Kroch Library Rare & Manuscripts Archives, Cornell University.

Saguisag, Rene. "There's a Jeepney in the Filipino Soul." *Mr. & Ms.* (Manila), October 30, 1984, 10.

Sanger, Frank W., and Electric Car Society. *Official Catalogue and Program of the First Annual Automobile Show.* Brooklyn, NY: Electric Car Society, 2003. CD-ROM.

Sass, Herman. *Willys 77: America's Depression Years Model T and Ancestor of the Jeep.* Buffalo: H. Sass, 1988.

The Saturday Evening Post. 1899, 1904, 1923.

"Says Autos Are Replacing Sleighs." *The Automobile*, February 13, 1908, 220.

Scharff, Virginia. *Taking the Wheel: Women and the Coming of the Motor Age.* New York: Free Press, 1991.

Schiffer, Michael B., Tamara C. Butts, and Kimberly K. Grimm. *Taking Charge: The Electric Automobile in America.* Washington, DC: Smithsonian Institution Press, 1994.

Schivelbusch, Wolfgang. *The Railway Journey: The Industrialization of Time and Space in the 19th Century.* Berkeley: University of California Press, 1986.

Schneider, Joseph. "Cultural Lag: What is it?" *American Sociological Review* 10, no. 6 (December 1945): 786–787.

Schncirov, Matthew. *The Dream of a New Social Order: Popular Magazines in America, 1893–1914.* New York: Columbia University Press, 1994.

Schreier, Konrad F., Jr. "Born for Battle." In Richards, *Military Jeeps 1941–1945*, 30–39.

———. "The Military Model T Ford." *Military Collector & Historian* 39, no. 3 (1987): 98–107.

Scott, Charles L., Colonel. "Progress in Cavalry Mechanization: Scout Car Developments." *The Cavalry Journal* 45, no. 4 (July–August 1936): 281–284.

Scott, Graham. *Essential Military Jeep: Willys, Ford & Bantam Models, 1941–45.* Biddeford, Devon, U.K.: Bay View Books, 1996.

Sejima, Hiro. From the Editor. *4X4 Magazine*, October 1999, 6–7.

"The Selection of a Gasoline Automobile According to Price." *The Automobile*, April 18, 1903, 422–434.

Sennett, A. R., and Electric Car Society. *Carriages Without Horses Shall Go.* Brooklyn, NY: Electric Car Society, 2003. CD-ROM.

Seton-Watson, Christopher. *Dunkirk, Alamein, Bologna: Letters and Diaries of an Artilleryman, 1939–1945.* London: Buckland, 1993.

"Severe Punishment to Break Up Joy Riding." *The Horseless Age*, December 8, 1909, 674.

Shimokawa, Kōichi. *The Japanese Automobile Industry: A Business History.* London; Atlantic Highlands, NJ: Athlone Press, 1994.

Shrader, Charles R. *United States Army Logistics, 1775–1992: An Anthology.* CMH Pub. Vol. 68. Washington DC: Center of Military History, United States Army, 1997.

Silverstone, Roger. *Media, Technology and Everyday Life in Europe: From Information to Communication.* Aldershott, U.K.: Ashgate Publishing Limited, 2005.

———. *Television and Everyday Life.* London: Routledge, 1994.

Sinclair, Harold. *The Cavalryman.* 1st ed. New York: Harper, 1958.

———. *The Horse Soldiers.* 1st ed. New York: Harper, 1956.

Smith, Stanley E. *The United States Marine Corps in World War II; the One-Volume History, from Wake to Tsingtao, by the Men Who Fought in the Pacific and by Distinguished Marine Experts, Authors, and Newspapermen*. New York: Random House, 1969.

Staff Correspondent. "The Sixth Cavalry in the Fourth Corps Maneuvers." *The Cavalry Journal* 49, no. 219 (May–June 1940): 194–200.

Stayer, Colonel Edgar S. "The Year's Advancement in Military Motor Transport." *The Quartermaster Review* 12 (July–August 1932): 33–37.

Steffen, Randy. *The Horse Soldier, 1776–1943: The United States Cavalryman: His Uniforms, Arms, Accoutrements, and Equipments. Vol. IV: World War I, the Peacetime Army, World War II: 1917–1943*. Norman, OK: University of Oklahoma Press, 1979.

Stettinius, Edward R. *Lend-Lease, Weapon for Victory*. New York: The Macmillan Company, 1944.

Stewart, Doug. "Hail to the Jeep! Could We Have Won Without It?" *Smithsonian* 23, no. 8 (November 1992): 60–73.

Stewart, H. S., Colonel. Mechanization and Motorization: The Final Chapter Has Not Been Written." *The Cavalry Journal* 49, no. 217 (January–February 1940): 35–41.

———. "Mechanized Cavalry Has Come to Stay." *The Cavalry Journal* 47, no. 6 (November–December 1938): 281–284.

Stodter, John Hughes, Major. "Radio Equipment for Horse Cavalry." *The Cavalry Journal* 50, no. 3 (May–June 1941): 68–69.

Street, Julian. "The Fools at the Finish." *Collier's*, November 7, 1908, 16–17.

Stubbs, Mary Lee, Stanley Russell Connor, and United States Dept. of the Army. Office of Military History. *Armor-Cavalry*. Army Lineage Series. Washington: U.S. Govt. Print. Off., 1969.

Surles, A. D., Major. "Cavalry Now and to Come." *The Cavalry Journal* 40, no. 164 (March–April 1931): 5–6, 64.

"Survival of the Horseless Carriage." *The Automobile*, August 21, 1906, 154.

Sweet, John Joseph Timothy. "Ferrea Mole, Ferreo Cuore: The Mechanization of the Italian Army, 1930–1940." PhD diss., Kansas State University, 1976.

Swinson, Arthur. *The Raiders: Desert Strike Force*. Ballantine's Illustrated History of World War II. Vol. 2. New York: Ballantine Books, 1968.

Taylor, James. *Jeep: CJ to Grand Cherokee*. Collector's Guide. Croydon, U.K.: Motor Racing Publications, 1999.

Taylor, Royce R. *The Cavalry and Armor Heritage Series*. Fort Knox, KY: United States Armor Association, 1986.

Thompson, F. M. L. "Victorian England: The Horse-drawn Society, An Inaugural Lecture." London: Bedford College, October 11, 1970.

Thurston, R. H. "The Automobile or Horseless Carriage." *Collier's*, April 28, 1900, 8–9.

———. "The Coming Automobile." *Collier's*, April 27, 1901, 9.

Tinio, Rolando S. "Romancing the Jeepney." *Man Magazine* (Manila), January 1993, 4.

Tobin, James. *Ernie Pyle's War: America's Eyewitness to World War II*. New York: Free Press, 1997.

Torre, Nestor, Jr. "Jeepney Soul." *The Philippines*, 1983, 48.

Torres, Emmanuel. *Jeepney*. Quezon City, Philippines: GCF Books, 1979.

———. "Numero Uno's Pamana: Legaspi Reinvents the Jeepney." *Sunday Globe Magazine* (Manila), September 27, 1988, 11.

Towle, Herbert L. "The Best Car for the Novice?" *The Automobile*, October 12, 1905, 395–397.

———. "The Coming of the Automobile." *Collier's*, January 12, 1901, 32–33.

Transportation: The Corps and the Function of the U.S. Army. Washington, DC: GPO, 1957.

Turner, Dawson. "Some Experiences with Modern Motor-Cars." *The Living Age Company*, June 9, 1900, 638–642.

"Two Auto Bills Pending in Michigan." *The Automobile*, January 21, 1905, 160.

Underwood, John. *Whatever Became of the Baby Austin?* El Monte, CA: R & L Press, 1965.

Union List of Serials in Libraries of the United States and Canada. Third Edition, Volume 1, A–B, edited by Edna Titus Brown, under the sponsorship of the Joint Committee on the Union List of Serials, with the cooperation of the Library of Congress. New York: H. W. Wilson Co., 1965.

United States Army. Cavalry. *Cavalry Life! In the U.S. Army "If You Want to Have a Good Time 'Jine' the Cavalry!" (Jeb Stuart's Song)* [U.S. Army Cavalry Recruiting Poster]. GPO, 1920. http://hdl.loc.gov/loc.pnp/cph.3g07550.

United States Cavalry Association. *The Cavalry Journal.* 1920–1946.

United States. Congress. House. Committee on Appropriations. Subcommittee on Deficiency Appropriations. *Defense Aid (Lend-Lease) Supplemental Appropriation Bill, 1943.* Washington, DC: GPO, 1973.

———. *Second Supplemental National Defense Appropriation Bill for 1942.* Washington, DC: GPO, 1973.

United States. Congress. House. Committee on Appropriations. Subcommittee on Deficiency Appropriations, United States. Congress. House. Committee on Appropriations. Subcommittee on War Department Appropriations, and United States. Congress. House. Committee on Appropriations. Subcommittee on Navy Department Appropriations. *Second Supplemental National Defense Appropriation Bill for 1942.* Washington, DC: GPO, 1973.

———. *Third Supplemental National Defense Appropriation Bill for 1942*. Washington, DC: GPO, 1973.

United States. Congress. House. Committee on Armed Services. *Subcommittee Hearings on H.R. 5342 and H.R. 5343, to Authorize the Secretary of Defense to Lend Certain Army, Navy, and Air Force Equipment to the Boy Scouts of America for Use at the Second National Jamboree of the Boy Scouts*. Washington, DC: GPO, 1973.

United States. Congress. House. Committee on Armed Services. Subcommittee No. 1. *Subcommittee Hearings on H.R. 4646, to Authorize the Secretary of the Army to Lend Certain Property of the Department of the Army to National Veterans' Organizations for Use at National Youth Tournaments*. Washington, DC: GPO, 1973.

United States. Congress. House. Committee on Foreign Affairs. *Extension of Lend-Lease Act*. Washington, DC: GPO, 1973.

United States. Congress. House. Committee on Foreign Affairs Subcommittee No. 2 (Foreign Affairs), and United States. Congress. House. Committee on International Relations. *Selected Executive Session Hearings of the Committee, 1943–50; Volume II: Problems of World War II and Its Aftermath*. Washington, DC: GPO, 1983.

United States. Congress. House. Committee on Foreign Affairs Subcommittee No. 6 (Foreign Affairs), and United States. Congress. House. Committee on International Relations. *Selected Executive Session Hearings of the Committee, 1943–50; Volume I: Problems of World War II and Its Aftermath*. Washington, DC: GPO, 1983.

United States. Congress. House. Committee on Foreign Affairs, and United States. Congress. House. Committee on International Relations. *Selected Executive Session Hearings of the Committee, 1943–50; Volume I: Problems of World War II and its Aftermath*. Washington, DC: GPO, 1983.

———. *Selected Executive Session Hearings of the Committee, 1943–50; Volume II: Problems of World War II and its Aftermath*. Washington, DC: GPO, 1983.

United States. Congress. House. Committee on Science, Space, and Technology. Subcommittee on Transportation, Aviation, and Materials. *Electric Vehicle Technology and Commercialization*. Washington, DC: GPO, 1991.

United States. Congress. House. *Lend-Lease Bill: Hearings Before the Committee on Foreign Affairs, House of Representatives, Seventy-Seventh Congress, First Session*. Washington, DC: GPO, 1941.

United States. Congress. House. *Subcommittee of the Committee on Public Buildings and Grounds, Hearings, Disposition of Surplus Property, Seventy-Eighth Congress, Second Session*. Washington, DC: GPO, 1943.

United States. Congress. Senate. Committee on Foreign Relations. *Lend-Lease*. Washington, DC: GPO, 1973.

United States. Congress. Senate. *Extension of the Lend-Lease Act, Hearings Before the Committee on Foreign Relations, United States Senate, Seventy-Eighth Congress, March 11, 1941*. Washington, DC: GPO, 1941.

United States. Congress. Senate. *Extension of the Lend-Lease Act, Hearings Before the Committee on Foreign Relations, United States Senate, Seventy-Eighth Congress, March 1, 1943*. Washington, DC: GPO, 1943.

United States. Congress. Senate. *Lend-Lease Aid, Preliminary Report of Committee Investigators to the Committee on Appropriations, United States Senate on Lend-Lease Aid and Government Expenditures Abroad, May 1, 1944*. Washington, DC: GPO, 1944.

United States. Congress. Senate. *Lend-Lease, Hearings Before the Committee on Foreign Relations, United States Senate, Seventy-Ninth Congress, March 28 and April 4, 1945*. Washington, DC: GPO, 1943.

United States. Congress, Senate, Foreign Trade Subcommittee, Special Committee to Study and Survey Problems of American Small Business, Hearings, *Problems of American Small Business*, 79th Cong., 1st Sess., 1945. Washington, DC: GPO, 1945.

United States. Office of Strategic Services. Research and Analysis Branch. *Summary of Strategic Information on Japan's Motor Truck Position*. Its R & A. Vol. 1743. Washington, DC: GPO, 1944.

United States. Philippine Commission, 1899–1900. *Report of the Philippine Commission to the Secretary of War*. 1900, 1915.

United States. Philippine Commission, 1899–1900, Jacob Gould Schurman, George Dewey, Elwell Stephen Otis, Charles Denby, and Dean C. Worcester. *Report of the Philippine Commission to the President, January 31, 1900[–December 20, 1900]*. 4 vols. Washington, DC: GPO, 1900, 1901.

United States. War Department, Office of the Chief of Staff. *Cavalry Drill Regulations, United States Army, 1916*. War Department Document No. 561. Washington, DC: GPO, 1917.

————. *Cavalry Service Regulations, United States Army (experimental), 1914*. War Department Document No. 461. Washington, DC: GPO, 1914.

————. *Drill Regulations for Cavalry, United States Army*. War Department Document No. 340. Washington, DC: GPO, 1909.

————. *Drill Regulations for Cavalry, United States Army: Amended 1909, Corrected to January 1, 1911*. War Department Document No. 340. Washington, DC: GPO, 1911.

————. *Drill Regulations for Machine-Gun Organizations, Cavalry, 1910 / [War Department, Office of the Chief of Staff]*. War Department Document No. 377. Washington, DC: GPO, 1910.

————. *Tentative Cavalry Drill Regulations, United States Army, November, 1913: School of the Platoon, Squadron, Regiment, and Brigade (Mounted Work Only)*. War Department Document No. 452. Washington, DC: GPO, 1913.

The Unstoppable Soldier. Produced by John V. Coscia and William Stephens, written by William Stephens. Global Television Network, ltd., 1996.

Ureta, Anjie Blardony. "The Jeepney is Still 'King of the Road.'" *Philippine Free Press*, October 12, 1991, 38.

Urlanda, Randy V. "Jeep and Jeepney Assemblers Cooperative." *Philippine Panorama*, April 24, 1994, 10.

———. "Jeep-Making: Imus' Sunshine Industry." *Philippine Panorama*, September 21, 1997, 12.

Urwin, Gregory J. W. *The United States Cavalry: An Illustrated History, 1776–1944*. Red River Books ed. Norman: University of Oklahoma Press, 2003.

———. *The United States Infantry: An Illustrated History, 1775–1918*. Norman: University of Oklahoma Press, 2000.

Van Creveld, Martin L. *Supplying War: Logistics from Wallenstein to Patton*. Cambridge; New York: Cambridge University Press, 1977.

Van Voorhis, Daniel, Colonel. "Packs and Leading." *The Cavalry Journal* 39, no. 161 (October 1930): 498–500.

Vidmer, George. "Major General William Ames Holbrook." *The Cavalry Journal* 41, no. 174 (November–December 1932): 41–42.

Wajcman, Judy. *Feminism Confronts Technology*. University Park: Pennsylvania State University Press, 1991.

Wakefield, Ernest Henry. *History of the Electric Automobile: Battery-Only Powered Cars*. Warrendale, PA: Society of Automotive Engineers, 1993.

Walker, Kirby. "Cavalry in the World War." *The Cavalry Journal* 33, no. 134 (January 1924): 11–22.

Well, A. Wade. *Hail to the Jeep*. New York: Harper & Brothers, 1946.

Wellman, Walter. "Faster than the Express-Train: The Automobile Race from Paris to Berlin." *McClure's Magazine*, November 1901–April 1902, 21–32.

White, George M., Lieutenant. "Cavalry's Iron Pony." *The Cavalry Journal* 50, no. 2 (March–April 1941): 85–88.

Whitehead, Arthur K., Major. "With the 26th Cavalry (P.S.) in the Philippines." *The Cavalry Journal* 53, no. 3 (May–June, 1944), 34–43.

"Why Not Here?" *Life*, August 17, 1905, 205.

Wik, Reynold. *Henry Ford and Grass-Roots America*. Ann Arbor: University of Michigan Press, 1972.

Willoughby, Charles Andrew, and John Chamberlain. *MacArthur, 1941–1951*. New York: McGraw-Hill, 1954.

———. *MacArthur: 1941–1951: Victory in the Pacific*. London: Heinemann, 1956.

Wilson, Arthur, Captain. "The Mechanized Force: Its Organization and Present Equipment," *The Cavalry Journal* 40, no. 165 (May–June 1931): 7–10.

Wilson, John B. *Maneuver and Firepower: The Evolution of Divisions and Separate Brigades*. Army Lineage Series. Washington, DC: Center of Military History, United States Army, 1998.

Winner, Langdon. *Autonomous Technology: Technics-out-of-Control as a Theme in Political Thought*. Cambridge, MA: MIT Press, 1978.

———. "Do Artifacts have Politics?" In *The Whale and the Reactor: A Search for Limits in an Age of High Technology*, edited by Langdon Winner, 19–39. Chicago: University of Chicago Press, 1986.

Wittgenstein, Ludwig. *Philosophical Investigations*.Translated by G. E. M. Anscombe. Oxford: Blackwell, 1997.

Woods, C. E. *The Electric Automobile: Its Construction, Care and Operation*. Chicago; New York: H. S. Stone & Company, 1900.

Woods, C. E., and the Electric Car Society. *The Electric Automobile: Its Construction, Care and Operation*. Brooklyn, NY: Electric Car Society, 2003. CD-ROM.

Woodward, B. D. "The Exposition of 1900," *The North American Review*, April 1900, 472–497.

Woolgar, Steve. "Configuring the User: The Case of Usability Trials." In *A Sociology of Monsters*, edited by John Law, 57–102. London: Routledge, 1991.

Woolgar, Steve, and Geoff Cooper. "Do Artifacts Have Ambivalence? Moses' Bridges, Winner's Bridges and other Urban Legends in S&TS." *Social Studies of Science* 29, no. 3 (1999): 433–449.

Wormser, Richard Edward. *The Yellowlegs: The Story of the United States Cavalry*. 1st ed. Garden City, NY: Doubleday, 1966.

Wright, Robert. *Nonzero: The Logic of Human Destiny*. New York: Pantheon Books, 2000.

Yale, Wesley W., First Lieutenant. "The Influence of Pack Loads on the Employment of Cavalry." *The Cavalry Journal* 43, no. 184 (July–August 1934): 19–21.

Yale, Wesley W., Colonel, General I. D. White, and General Hasso E. von Manteuffel. *Alternative to Armageddon*. New Brunswick, NJ: Rutgers University Press, 1970.

Yasuo, Ōtsuka. *Ōtsuka Yasuo no Omochabako* [Ōtsuka Yasuo's Toybox]. Self-published, 1996.

Yin, Robert K. *Case Study Research: Design and Methods*. Thousand Oaks, CA: Sage Publications, 1994.

Young, Peter. *The World Almanac of World War II: The Complete and Comprehensive Documentary of World War II*. 1st rev. ed. New York: World Almanac, 1986.

Young, Theresa Dwyre. "Should Off-Road Vehicles be Allowed in National Parks?" *Backpacker*, June 2005, 48.

Young, Walter E., and John J. Vander Velde. *Cavalry Journal/Armor Cumulative Indices, 1888–1968*. Bibliography Series—Kansas State University

Library; no. 12. Manhattan, KS: Kansas State University Library, 1974.

"Youthful Pacific Coast Driver." *The Automobile*, February 22, 1906, 425.

Zeck, Judith, and United States International Trade Commission. *Certain All-Terrain Vehicles from Japan: Determination of the Commission in Investigation no. 731-TA-388 (Preliminary) Under the Tariff Act of 1930, Together with the Information Obtained in the Investigation.* USITC Publication 2071. Washington, DC: US International Trade Commission, 1988.

Zeichner, Walter. *Jeep: Willy's, Kaiser, AMC, 1942–1986: A Documentation.* Schiffer Automotive Series. West Chester, PA: Schiffer Pub, 1990.

Zipper, Trudl Dubsky, and Herbert Zipper. *Manila 1944–45 as Trudl Saw It: Watercolors of Trudl Dubsky Zipper.* Santa Monica, CA: Crossroads School in cooperation with Herbert Zipper, 1994.

INDEX